Collins · *do brilliantly!*

InstantRevision

AS Physics

Contents

Published by HarperCollins*Publishers* Ltd
77-85 Fulham Palace Road
London W6 8JB

www.**Collins**Education.com
On-line support for schools and colleges

© HarperCollins*Publishers* 2002

First published 2002
This new format edition published 2004
10 9 8 7 6 5 4
ISBN-13 978 0 00 717268 9
ISBN-10 0 00 717268 0

Martin Gregory asserts the moral right to be identified as the author of this work.

British Library Cataloguing in Publication Data
A catalogue record for this book is available from the British Library.

Edited by Jane Bryant
Production by Katie Butler
Design by Gecko Ltd
Illustrations by Gecko Ltd
Cover design by Susi Martin-Taylor
Printed and bound by Printing Express Ltd, Hong Kong

Every effort has been made to contact the holders of copyright material, but if any have been inadvertently overlooked, the Publishers will be pleased to make the necessary arrangements at the first opportunity.

Get the most out of your Instant Revision pocket book

1 **Maximise your revision time.** You can carry this book around with you anywhere. This means you can spend any spare moments revising.

2 **Learn and remember what you need to know.** The book contains all the really important facts you need to know for your exam. All the information is set out clearly and concisely, making it easy for you to revise.

3 **Find out what you don't know.** The *Check yourself* questions help you to discover quickly and easily the topics you're good at and those you're not so good at.

What's in this book

1 The content you need for *your* AS exam

- This book covers the six AS Physics specifications of the three Awarding Bodies (AQA, Edexcel and OCR). Each Awarding Body has two Physics specifications: AQA (A) and AQA (B); Edexcel and Edexcel (SH); OCR (A) and OCR (B). Make sure you know which exam you are entered for as the content requirements differ between the six specifications.

- Where whole topics are only required for certain specifications, we have indicated this clearly.

2 Formulae you must know

- Physics is a quantitative subject. Physicists put numbers into their explanations, so it is important that you feel confident in handling straightforward mathematics and performing routine calculations. You are expected to know certain formulae; they will not be given to you in the exam or on the formulae sheet. The formulae you need to know are highlighted in the text and are listed on page 120.

3 *Check yourself* questions – find out how much you know and improve your grade

- The *Check yourself* questions appear at the end of each short topic.

- The questions are quick to answer. They are not actual exam questions, but the author is an examiner and has written them in such a way that they will highlight any vital gaps in your knowledge and understanding.

- The answers are given at the back of the book. When you have answered the questions, check your answers with those given. The author gives help with arriving at the correct answer, so if your answer is incorrect, you will know where you went wrong.

- There are marks for each question. If you score very low marks for a particular *Check yourself* page, this shows that you are weak on that topic and need to put in more revision time.

Revise actively!

- **Concentrated, active revision** is much more successful than spending long periods reading through notes with half your mind on something else.

- The chapters in this book are quite short. For each of your revision sessions, choose a couple of topics and concentrate on reading and thinking them through for **20–30 minutes**. Then do the *Check yourself* questions. If you get a number of questions wrong, you will need to return to the topic at a later date. Some Physics topics are hard to grasp but, by coming back to them several times you will become more confident about the exam.

- Use this book to revise either on your own or with a friend!

Quantities and units

A physical quantity such as a *mass of 2 kg*, has a **magnitude** (in this case the number 2) and a unit (in this case kg). **Always write the unit with the magnitude of a quantity**, especially when you are doing calculations. In any equation the units on each side must be the same and, by writing in the units for each quantity, you have a check that you are using the correct equation in the correct manner.

For example:

final speed $\quad=\quad$ initial speed $\quad+\quad$ (acceleration × time)

$12\,\text{m}\,\text{s}^{-1}\quad=\quad 4\,\text{m}\,\text{s}^{-1}\quad+\quad(2\,\text{m}\,\text{s}^{-2} \times 4\,\text{s})$

The units on each side are the same ($\text{m}\,\text{s}^{-1}$ for each term or group of terms). By writing in the units we have shown we are using the correct equation for a motion calculation. Writing simply

$12\qquad=\qquad 4\qquad\qquad+\qquad (2 \times 4)$

is correct arithmetic but not correct physics. It could have referred to a different problem, such as:

$12\,\text{V}\qquad=\qquad 4\,\text{V}\qquad\qquad+\qquad (2\,\text{A} \times 4\,\Omega)$

SI units

All the units used in your physics course are **SI** (Système International) **units**. The system is based on seven **base** units, of which you use the following six:

- length (**metre**)
- mass (**kilogram**)
- time (**second**)
- temperature (**kelvin**)
- electric current (**ampère**)
- amount of substance (**mole**)

From these, a set of simply related **derived** units are defined: for example, a force of 1 **newton** accelerates 1 **kilogram** at 1 **metre per second**². Make sure you know how the appropriate units are defined. The newton is a derived unit defined as a $\text{kg}\,\text{m}\,\text{s}^{-2}$.

Significant figures

Real physical data have uncertainties associated with them. The number of **significant figures** (sig. figs) to which data are quoted tells you how accurate they are. For example:

- 4 (1 sig. fig.) means 'between 3 and 5'
- 4.0 (2 sig. figs) means 'between 3.9 and 4.1'
- 4.00 (3 sig. figs) means 'between 3.99 and 4.01'.

The result of a calculation cannot be more accurate than the least accurate data used. So, when data in a question are given to 2 sig. figs you should give your answer to 2 sig. figs. For example, in the question 'a cyclist moves 23 m in 7.5 s. Calculate his speed' the data are given to 2 sig. figs. Your answer should be $\frac{23\,\text{m}}{7.5\,\text{s}} = 3.1\,\text{m}\,\text{s}^{-1}$ – quoted to 2 sig. figs (not $3.0666667\,\text{m}\,\text{s}^{-1}$, which is what you will see on the calculator display).

Remember that you could be penalised in examinations for not giving your answer to the same number of sig. figs as the data in the question.

Mathematics you need to know

- You need to be able to work with numbers in standard form and with metric prefixes. Make sure you can convert data from one form to another. For example, the radius of the Earth is about

$6\,400\,000\,\text{m}$	(2 sig. figs in longhand)
$= 6\,400\,\text{km}$	(2 sig. figs using the prefix k)
$= 6.4 \times 10^6\,\text{m}$	(2 sig. figs in standard form)

Make sure you know how to enter standard form on your calculator.

Operation	Prefix	Standard form
multiply by one billion	giga (G)	$\times 10^9$
multiply by one million	mega (M)	$\times 10^6$
multiply by one thousand	kilo (k)	$\times 10^3$
divide by one thousand	milli (m)	$\times 10^{-3}$
divide by one million	micro (µ)	$\times 10^{-6}$
divide by one billion	nano (n)	$\times 10^{-9}$

- Metric prefixes you need to know are shown on page 2.
- You need to be able to work out ratios and percentages.
- In trigonometry, remember Pythagoras' theorem for a right–angled triangle: $a^2 + b^2 = c^2$.

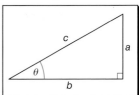

- Make sure you can use your calculator to find $\sin \theta = \frac{a}{c}$, $\cos \theta = \frac{b}{c}$ and $\tan \theta = \frac{a}{b}$ where θ is in degrees.
- Remember the formulae to work out:

the area of a triangle	=	$\frac{1}{2}$ base × height
the area of a circle	=	$\pi \times (\text{radius})^2$
the circumference of a circle	=	$2\pi (\text{radius})$
the surface area of a sphere	=	$4\pi (\text{radius})^2$
the volume of a sphere	=	$\frac{4\pi}{3} (\text{radius})^3$
the volume of a cylinder	=	$\pi \times (\text{radius})^2 \times (\text{height})$

- Use the notation 'change in x' $= x_2 - x_1 = \Delta x$.
- Make sure you can use your calculator to work out simple functions such as $x^2, \sqrt{x}, x^n, \frac{1}{x}, \log_{10}x, \log_e x$.
- When solving problems you often need to rearrange the subject of an equation before substituting numbers in it.

  For example: to work out the speed at which a mass with known kinetic energy is moving we need to rearrange the equation

  $$E_k = \frac{1}{2} m v^2$$

  to make v the subject.

  First, find v^2: $\quad v^2 = 2\dfrac{E_k}{m}$

  Then take the square root: $v = \sqrt{2\dfrac{E_k}{m}}$

- Make sure you know the rules for manipulating simple equations. For example:

  $y = 2x(3 + 2x)$ gives $y = 6x + 4x^2$ not $y = 6x + 2x$.

Graphs

When analysing a graph the first question in your mind should be 'is it a straight line or a curve?'

● If the graph is a **straight line** the quantities plotted as x and y are **proportional** when the graph goes **through the origin**, and have a **linear relationship** when the graph misses the origin.

● The equation for a straight-line graph is $y = mx + c$ where m is the **gradient** or slope of the graph and c is the **intercept** or the value of y where the line of the graph crosses the y axis. Find the gradient of the graph by measuring the change in y (Δy) divided by the change in x (Δx).

● Because straight-line graphs are the easiest to analyse it is often desirable to rearrange equations to fit $y = mx + c$ and obtain a straight line rather than a curved line. For example:

> plotting data to fit the equation $v^2 = 2gs$ in a motion experiment would give a curve (a parabola) when v is plotted against s but a straight line when v^2 is plotted against s ($y = v^2$, $x = s$, $m = 2g$, $c = 0$).

● To find the gradient when the graph is a curve **draw a tangent** to the curve at the point of interest and calculate the gradient of the tangent.

● In many graphs the area between the line and the x axis has significance (e.g. in a speed–time graph, the area between the line and the time axis gives the distance travelled). Make sure you can work out this area by calculation or by counting squares on the graph paper.

● You should be able to recognise the shapes of common types of graph:

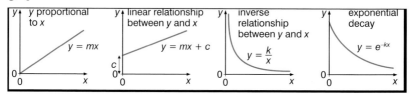

Check yourself

Units and mathematics

1 Write down the following quantities in standard form:
  (a) 2500 m (1)
  (b) 4.5 MJ (1)
  (c) 0.25 mA (1)

2 Express the following units in terms of the base units: m, kg, s.
  (a) the unit of energy, the joule (J) (2)
  (b) the unit of pressure, the pascal (Pa) (2)

3 Write down the answers to the following calculations to 3 sig. figs.
  (a) $\frac{2}{3}$ (1)
  (b) $\pi = 3.14159$ (1)
  (c) $(124 \times 153) + 1.67 \times 10^4$ (1)

4 For the angle θ in the triangle, calculate:
  (a) $\sin \theta$ (1)
  (b) $\cos \theta$ (1)
  (c) $\tan \theta$ (1)

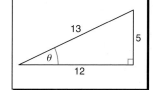

5 Rearrange the following formulae:
  (a) $V = IR$ to make I the subject (1)
  (b) $s = \frac{1}{2} gt^2$ to make t the subject (2)
  (c) $y = mx + c$ to make x the subject (1)

6 Combine the equations $V = IR$ and $R = \frac{\rho l}{A}$ to express ρ in terms of V, I, l and A by eliminating R. (3)

The answers are on page 102.

Vectors and scalars

All physical quantities can be classified as vectors or scalars. Physical quantities that have a direction associated with them are **vectors**. Quantities that are not associated with particular directions are called **scalars**. For instance, distance travelled is a **scalar** quantity because it has **magnitude only** (the direction is not specified) but displacement is a **vector** quantity because it has **both magnitude and direction**.

● **Scalar quantities** include mass, distance, speed, energy, work, power.

● **Vector quantities** include displacement, velocity, acceleration, force, weight, momentum.

Scalar quantities can be added, subtracted, etc. using the rules of arithmetic.

When combining **vector** quantities take notice of their directional properties. This can be done graphically or by calculation. The vectors are represented by arrows showing their direction.

● To combine or add graphically the two vectors **OA** and **OB** acting at O, draw a scale diagram with the two vectors **OA** and **OB** represented by arrows of length proportional to the magnitude of the 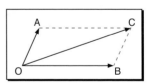 vectors and in the appropriate directions. The **resultant vector** is given by the diagonal **OC**. From the length of **OC** you can find the magnitude of the resultant, and its direction is that of **OC**.

● When the two vectors a and b are perpendicular to each other you can calculate the resultant vector c using Pythagoras' theorem $(a^2 + b^2 = c^2)$.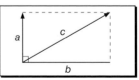

Worked example

Find the ground speed of an aircraft flying due North at $100\,\mathrm{m\,s^{-1}}$ in air with a wind speed of $40\,\mathrm{m\,s^{-1}}$ blowing from West to East.

Draw a vector diagram to scale (e.g. $10\,\mathrm{m\,s^{-1}} = 10\,\mathrm{mm}$) and find the length of the diagonal, which is the ground speed. On this scale, length of diagonal = 110 mm corresponding to a speed of $110\,\mathrm{m\,s^{-1}}$.

Alternatively, calculate using Pythagoras' theorem:

Ground speed $=\sqrt{[(100\,\mathrm{m\,s^{-1}})^2 + (40\,\mathrm{m\,s^{-1}})^2]}$
$= 108\,\mathrm{m\,s^{-1}}$.

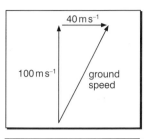

Vectors can be **resolved** into two **perpendicular component vectors**. For example; the weight (a force vector downwards) of a car going down a hill can be resolved into a component vector parallel to the road (accelerating the car down the hill) and a component vector perpendicular to the road (the force exerted by the wheels on the road).

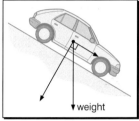

Resolution of vectors is a useful technique because each component vector may be regarded as an independent quantity and the problem solved separately for each component in turn.

In the diagram the vector OA has components OA cos θ in the OX direction and OA sin θ in the OY direction. All the components in a particular direction can be added arithmetically.

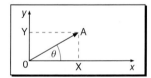

Velocity

When an object moves we often want to work out its **speed**.
- **average speed = distance travelled divided by time taken**
$$v = \frac{s}{t}$$

It is worth remembering this simple formula in all its forms:
- **distance travelled = average speed multiplied by time taken**
- **time taken = distance travelled divided by average speed.**

Worked example:

An athlete runs a 400 m race in 45 s. Calculate his average speed.

Average speed $= \dfrac{\text{distance}}{\text{time}} = \dfrac{400\,\text{m}}{45\,\text{s}} = 8.9\,\text{m s}^{-1}$.

In most of the problems you will come across in AS Physics the object is moving in a straight line, so that the distance travelled in the given direction is the **displacement** of the object from its starting or reference point. The speed of the object is also in the direction of the displacement and is called the **velocity**. Both displacement and velocity are vector quantities.

● velocity = displacement divided by time

$$v = \frac{s}{t} \ \text{ or } \ \frac{\Delta s}{\Delta t}$$

Here Δs is a small change in displacement taking place in time Δt. This way of writing velocity enables you to consider situations where the velocity is not constant so there is a different Δs for each Δt.

● When the velocity is positive the displacement is increasing; the body is moving away from its starting point.
● When the velocity is negative the body is moving to reduce its displacement.
● When the velocity increases the body accelerates, when it decreases the body decelerates.

Acceleration

Acceleration is the rate of change of velocity with time:

● acceleration = change in velocity divided by time taken for change.

$$a = \frac{(v - u)}{t} = \frac{\Delta v}{\Delta t}$$

where u is initial velocity, v is the final velocity and t is the time taken for the velocity to change.

Note that if the final velocity is less than the initial velocity the acceleration is negative. **Deceleration** is **negative acceleration**.

Units

Velocity and speed are both measured in $m\,s^{-1}$ and acceleration in $m\,s^{-2}$. Remember to change velocities and accelerations into these SI units before using them in calculations.

Worked example

A car accelerates from rest to $50\,km\,h^{-1}$ in 11 s. Calculate its acceleration.

First convert $km\,h^{-1}$ to the correct SI units ($m\,s^{-1}$). $50\,km = 50\,000$ m, $1\,h = 3600\,s$.

Therefore $50\,km\,h^{-1} = \frac{50\,000\,m}{3600\,s} = 14\,m\,s^{-1}$.

Acceleration $= \frac{(v-u)}{t} = \frac{(14\,m\,s^{-1} - 0)}{11\,s} = 1.3\,m\,s^{-2}$.

Distance/time graphs

On the distance/time graph:
- OA represents the total distance travelled in time t.
- The gradient of the graph represents the speed.

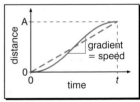

On this graph the speed varies (the graph line is not a straight line). The gradient of the dotted line gives the average speed for the journey.

Displacement/time graphs

On this graph:
- in section OA the gradient is increasing (acceleration)
- in section AB the gradient is constant (constant velocity)
- in section BC the displacement does not change (the body is at rest)
- in section CD the object returns to its starting point (there is zero displacement at D).

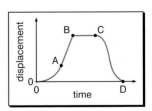

Velocity/time graphs

On the velocity/time graph:
- the gradient of the graph gives the acceleration
- the shaded area between the graph and the time axis gives the displacement.

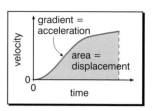

Worked example

The graph shows how the speed of a runner at the start of a 100 m sprint race varies with time. Find (a) his acceleration over the first 4.0 s, and (b) the distance covered in the first 4.0 s.

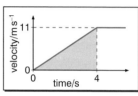

(a) His acceleration is the gradient of the graph over the first 4.0 s, which equals $\dfrac{11\,\mathrm{m\,s^{-1}}}{4.0\,s} = 2.8\ \mathrm{m\,s^{-2}}$.

(b) The area under the part of the graph during which he is accelerating is the area of the shaded triangle.

Area $= \frac{1}{2}(\text{base}) \times (\text{height})$

Therefore distance travelled $= \frac{1}{2}(4.0\,\mathrm{s}) \times (11\,\mathrm{m\,s^{-1}}) = 22\,\mathrm{m}$.

He accelerates at $2.8\,\mathrm{m\,s^{-2}}$ and covers 22 m in 4.0 s.

Check yourself

Velocity, acceleration and motion

1 Explain how two vectors of magnitudes 1 and 2 can be combined to give resultant vectors with magnitudes of 1 and 3. Draw appropriate vector diagrams. (1)

2 A swimmer has a speed in the water of $1.2\,m\,s^{-1}$. He swims across a river flowing at $0.50\,m\,s^{-1}$. By drawing an appropriate scale diagram find his ground velocity when he swims perpendicular to the river flow. (3)

3 Calculate the average speed, in $m\,s^{-1}$, of a train that goes from London to Winchester (105 km) in one hour. (2)

4 Measurements on a stone dropped down a well show that it was moving at $10\,m\,s^{-1}$ one second after release from rest. Calculate a value for the acceleration of its fall. (2)

5 A motor car accelerates from rest to $30\,m\,s^{-1}$ in 6.0 s. Calculate its average acceleration. (2)

6 A racing car completes one lap of a circuit of length 4000 m in 90 s. Calculate:
   (a) Its average speed. (1)
   (b) Its average velocity over a complete lap. Explain your reasoning. (2)

7 Here is a displacement/time graph for a journey.
   (a) Describe the journey in as much detail as you can. (3)
   (b) Draw a velocity/time graph for the journey. (3)

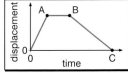

8 Draw a velocity/time graph for a truck which accelerates at $2.5\,m\,s^{-2}$ for 8.0 s and then slows to a stop in a further 15 s. (4)

9 The graph shows the variation of velocity with time for the first 40 s of a journey. Estimate the distance travelled. (3)

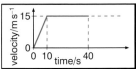

The answers are on page 102.

If we do not have a velocity/time graph we can calculate the quantities a and s algebraically.

For motion in a **straight line**, either at constant velocity or at constant acceleration, we can use four equations to relate the distance travelled s, the initial velocity u, the final velocity v, the acceleration a and the time t. These equations are worth remembering, even though they are given on formulae sheets, because they appear in so many problems in a wide range of topics. The four equations of motion are:

- $v = u + at$ [1]

- $s = \frac{1}{2}(u + v)t$ [2]

- $s = ut + \frac{1}{2}a\,t^2$ [3]

- $v^2 = u^2 + 2as$ [4]

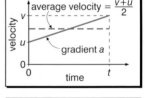

Equation [1] comes from rearranging the definition of acceleration from page 9:

$$a = \frac{(v - u)}{t}.$$

Equation [2] is just 'distance = average speed multiplied by time'

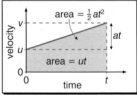

Equation [3] comes from the formulae used to calculate the area between the velocity/time graph line and the time axis.

Equation [4] comes from combining equations [1] and [3].

Rearrange equation [1] to give $t = \frac{(v - u)}{a}$ and substitute for t in equation [3]:

$$s = u\,\frac{(v - u)}{a} + \frac{1}{2}\frac{(v - u)^2}{a}$$

$2as = 2(uv - u^2) + (v^2 - 2uv + u^2) = v^2 - u^2$, which is equation [4].

You do not need to remember these derivations.

Worked examples

1 A train travelling at $20\,\mathrm{m\,s^{-1}}$ accelerates to $30\,\mathrm{m\,s^{-1}}$ in $15\,\mathrm{s}$. Calculate (a) the acceleration of the train, and (b) how far it travels whilst accelerating.

(a) The formula required to find the acceleration is $v = u + at$. The data given are $u = 20\,\mathrm{m\,s^{-1}}$, $v = 30\,\mathrm{m\,s^{-1}}$, $t = 15\,\mathrm{s}$. Rearranging the formula gives $a = \dfrac{v - u}{t}$

Substituting in the formula gives

$$a = \frac{30\,m\,s^{-1} - 20\,m\,s^{-1}}{15\,s} = 0.67\,\mathrm{m\,s^{-2}}.$$

(b) The formula required to find the distance is $s = \frac{1}{2}(u + v)t$

The data given are $u = 20\,\mathrm{m\,s^{-1}}$, $v = 30\,\mathrm{m\,s^{-1}}$, $t = 15\,\mathrm{s}$.

Substituting in the formula gives:

$$s = \tfrac{1}{2}(20\,\mathrm{m\,s^{-1}} + 30\,\mathrm{m\,s^{-1}}) \times 15\,\mathrm{s} = 375\,\mathrm{m}.$$

Therefore, the train accelerates at $0.67\,\mathrm{m\,s^{-2}}$ and travels $375\,\mathrm{m}$ during the acceleration.

2 The train, travelling at $30\,\mathrm{m\,s^{-1}}$, applies the brakes. It decelerates at $2.0\,\mathrm{m\,s^{-2}}$ to come to rest in a station. Calculate how far it travels in stopping.

The formula required to find the distance is $v^2 = u^2 + 2as$. The data given are $v = 0$, $u = 30\,\mathrm{m\,s^{-1}}$, $a = -2.0\,\mathrm{m\,s^{-2}}$. Rearranging the formula gives $s = \dfrac{-u^2}{2a}$.

Substituting in the formula gives $s = \dfrac{-(30\,\mathrm{m\,s^{-1}})^2}{2 \times -2.0\,\mathrm{m\,s^{-2}}} = 225\,\mathrm{m}.$

Therefore, the train comes to rest $225\,\mathrm{m}$ after applying the brakes.

Projectiles

When a ball is thrown into the air it follows a curved path. To work out its path we can make use of a very important simplification of the problem – the resolution of vectors – since velocity is a vector.

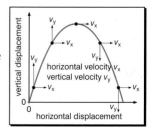

We can treat the problem as two linear motion problems at right angles to each other: the horizontal motion of the ball and the vertical motion of the ball. The two motions are said to be **independent**.

- In the **horizontal** direction, in the absence of frictional forces such as air resistance, the ball moves with **constant velocity** because no force acts on it in this direction.
- In the **vertical** direction, in the absence of frictional forces such as air resistance, the ball moves with **constant acceleration**, the acceleration due to gravity.

When we have solved these two separate problems using the equations of motion in a straight line, combining the two solutions enables us to work out the path of the ball.

Worked example

A golf ball is projected horizontally at 5.0 m s⁻¹ off a bench 0.90 m from the floor. Find how far from the edge of the bench the ball lands on the floor.

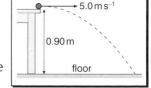

First find how long the ball takes to reach the floor from the **vertical** motion of the ball falling freely from rest under gravity:

The formula required is $s = ut + \frac{1}{2}gt^2$.

In this problem u is zero (falls from rest) so the formula rearranges to $t = \sqrt{\frac{2s}{g}}$.

The data given are $s = 0.90\,\text{m}$, $g = 9.8\,\text{m s}^{-2}$ (the acceleration due to gravity).

Substituting in the formula gives $t = \sqrt{\dfrac{2 \times 0.90\,\text{m}}{9.8\,\text{m s}^{-2}}} = 0.43\,\text{s}$.

Next use this answer to find how far the ball travels **horizontally**:

The formula required is $s = vt$.

The data given are $v = 5.0\,\text{m s}^{-1}$, $t = 0.43\,\text{s}$ (from first part of calculation).

Substituting in the formula gives $s = 5.0\,\text{m s}^{-1} \times 0.43\,\text{s} = 2.1\,\text{m}$.

Final answer: The ball lands 2.1 m from the edge of the bench.

The equations of motion

1 An electron in a cathode ray tube accelerates from rest to $2.5 \times 10^6 \, \text{m s}^{-1}$ in a distance of 35 mm. Calculate:

   (a) Its acceleration. (2)

   (b) How long it takes to reach the screen 0.40 m away. (1)

2 A car rolls from rest down a hill, accelerating at $0.15 \, \text{m s}^{-2}$.

   (a) Calculate how fast it is travelling when it has moved 75 m down the hill. (3)

   (b) How long does it take to travel the 75 m? (2)

3 A stolen car is travelling at a steady $30 \, \text{m s}^{-1}$ along a motorway. As it passes a police car, the police car accelerates from rest at $4.0 \, \text{m s}^{-2}$ to reach the speed of $36 \, \text{m s}^{-1}$, at which it chases the stolen car.

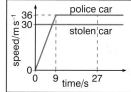

The graph shows the speed of both cars plotted against time. Show that the police car catches the stolen car after 27 s, having covered 810 m. (5)

4 A gun fires a shell with an initial velocity of $300 \, \text{m s}^{-1}$. Neglect air resistance in your calculations. The gun is fired vertically upwards. Calculate how long the shell takes:

   (a) To reach the highest point on its trajectory. (2)

   (b) To return from that point to the ground. (2)

5 The gun in question 4 is fired at 45° to the horizontal.

   (a) Calculate the component velocities in the vertical and horizontal directions. (2)

   (b) Using the vertical component velocity, calculate the time taken for the shell to rise to the top of its trajectory and fall back to Earth. (3)

   (c) Using the horizontal component velocity find the horizontal distance travelled while the shell goes up and down (the time calculated in **(b)**) and hence find where it lands. (3)

   (d) State the direction in which the shell is travelling at the top of its trajectory. (1)

The answers are on pages 103–104.

Forces

Forces are pushes and pulls. **Newton's first law** of motion identifies a force as that which changes the state of rest or uniform motion of a body.

● Forces arise from gravitational, electric, magnetic and nuclear interactions.
● When there is **zero resultant force** on a body it stays **at rest** or **moves with constant velocity**.
● Force is measured in **newtons (N)**. From Newton's second law of motion, $F = ma$, **1 N accelerates 1 kg at 1 m s^{-2}**.

Worked example

A catapult applies a force of 20 N to a stone of mass 0.15 kg. Calculate the initial acceleration of the stone as it leaves the catapult.

Use $a = \dfrac{F}{m}$

Here $F = 20$ N, $m = 0.15$ kg,

$a = \dfrac{20\,\text{N}}{0.15\,kg} = 130\,\text{m s}^{-2}$.

● Force is a vector quantity. The resultant of two or more forces must be found using the rules for combining vectors. The diagram shows forces of 3 N and 4 N combined to give resultant forces of 7 N, 1 N and 5 N.

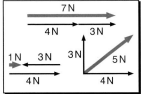

● There are two forces acting on this student: his **weight** acting downwards and the **reaction** force from the chair acting upwards. Such a diagram is called a **free-body force diagram** for the student.

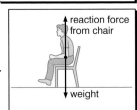

● The mass m in the equation $F = ma$ is called the **inertial mass**. It is that property of the body which relates the force to the acceleration produced. We make it equal to the quantity of matter mass defined by the kilogram (see below).

Mass, density and weight

Mass is a measure of the quantity of material in a body. It is measured in **kilograms (kg)** and is a scalar quantity.

The **mass per unit volume of a material** is called the **density** (ρ),

$$\rho = \frac{m}{V}$$

- Density is a characteristic of the material making up the body.
- Density is measured in $\mathbf{kg\,m^{-3}}$. If you come across density data given in $g\,cm^{-3}$, remember that:

$$1\,g\,cm^{-3} = 1000\ kg\,m^{-3}.$$

The **gravitational force** on a body is called its **weight**. It is measured in **newtons** and is a **vector** quantity. Your weight is the gravitational force of attraction between you and the Earth and acts on you towards the centre of the Earth.

- Weight = mass × gravitational field strength, $W = mg$.
- The gravitational field strength, g, near the surface of the Earth is $9.8\,N\,kg^{-1}$. From $F = ma$ you can check that $N\,kg^{-1}$ is the same as $m\,s^{-2}$. Thus g is also called the acceleration due to gravity.

Weight is a gravitational force and is only experienced in a gravitational field.

A person of mass 65 kg has a weight of $(65\ kg) \times (9.8\,N\,kg^{-1}) = 640\,N$ on the surface of the Earth.

On the surface of the Moon, where the gravitational field strength is $1.7\,N\,kg^{-1}$, the same person would have a weight of $(65\,kg) \times (1.7\,N\,kg^{-1}) = 110\,N$.

Astronauts in orbit are said to be 'weightless'. They still experience a gravitational force towards the Earth but the resultant force on them is zero because this gravitational force provides the centripetal force to keep them in orbit. They are effectively in 'free fall'.

The acceleration due to gravity

When you drop a ball so that it falls to the floor, the ball accelerates as it falls. In the absence of air resistance all bodies fall with the same acceleration near the surface of the Earth. Remember Galileo's famous experiment at the leaning tower of Pisa!

● The acceleration of free fall near the Earth's surface is about 9.8 m s^{-2} and is called the **acceleration due to gravity**, g.

The acceleration due to gravity is often measured in the school laboratory by releasing a steel ball from an electromagnet and timing its fall over a measured distance. The effects of air resistance are assumed to be negligible.

Worked example

Below is a table of data from a school experiment to measure acceleration due to gravity. What value will the results give?

distance, s, fallen/m	0.50	0.75	1.00	1.50
time, t to fall/s	0.32	0.40	0.45	0.55
calculated value of t^2/s^2	0.10	0.16	0.20	0.30

The equation needed is $s = \frac{1}{2} at^2$. In this case the acceleration $a = g$.

Calculate values of t^2 and plot a graph of s against t^2. The equation suggests that s is proportional to t^2 and the graph has a gradient of $\frac{1}{2} g$.

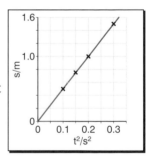

Looking at the graph you can see that s is proportional to t^2. The gradient is 5.0 m s^{-2}. This gives a value of
$g = 2 \times$ gradient $= 10.0$ m s^{-2}.

The measured acceleration due to gravity is 10.0 m s^{-2}.

Forces and moments of forces

Forces can produce rotation as well as linear motion. When you push on a door the door rotates about its hinges rather than moving forward in a straight line. The turning effect of a force is called the **moment of the force** or a **torque**.

● The moment of a force is greater the further the point of application of the force is from the hinge or pivot.

● The moment of a force or torque is the **magnitude of the force multiplied by the perpendicular distance of the line of action of the force from the pivot or centre of rotation**.

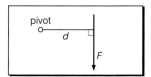
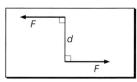

● The moment of a force or torque $= F \times d$.

● The units of torque are **newton metre (N m)**.

● Two equal forces whose lines of action do not coincide constitute a **couple**, Fd, which will make a body rotate.

For example, to tighten a nut and bolt to a torque of 15 N m you could use:

● a force of 15 N at the end of a spanner of length 1.0 m,

● a force of 150 N at the end of a 10 cm long spanner, or

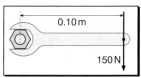

● a force of 1500 N at the end of a 1.0 cm spanner.

All three of these would exert the same torque or have the same moment.

Equilibrium

For a body to be in equilibrium there must be zero resultant force on it – otherwise it will move in the direction of the resultant force!

However, even though it may not move in a straight line, it may rotate if there are unbalanced moments of the forces acting.

● For equilibrium the resultant force **and** resultant torque, or moment of the forces, or couple, must be **zero**.

- In equilibrium, the **sum of the clockwise moments** about any point **equals the sum of the anticlockwise moments** about that point.

Worked example

Two children A and B with weights of 400 N and 450 N respectively each sit 1.5 m from the pivot of a see-saw. The see-saw does not balance. Where must B move to make the see-saw balance?

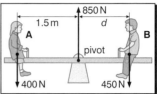

To balance the see-saw B needs to move to a distance d from the pivot. Balancing the moments gives:

total clockwise moment = total anticlockwise moment

$$(450\,\text{N})(d) = (400\,\text{N})(1.5\,\text{m})$$

$$d = \frac{(400\,\text{N} \times 1.5\,\text{m})}{450\,\text{N}} = 1.3\,\text{m}.$$

Therefore, B must move until he is 1.3 m from the pivot.

Centre of gravity

Real bodies have complex shapes and gravity acts on all parts of the body. All bodies have one point, called the **centre of gravity**, about which all the moments of the weights of the component elements balance out. When it is suspended from its centre of gravity a body does not try to rotate or twist. The **centre of mass** is exactly the same position – provided the gravitational field strength doesn't vary within the body.

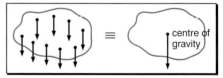

When solving problems we usually make use of the simplification that the body may be considered as a point mass whose weight is concentrated at its centre of gravity. We then put the 'weight force' arrow through the centre of gravity.

- The centre of gravity of a sphere is at its centre.
- The centre of gravity of a circular disc is at its centre.
- The centre of gravity of a uniform beam or rod (e.g. a metre rule) is at its mid-point.

Check yourself

Forces and moments of forces

For all these questions, assume the acceleration due to gravity, $g = 9.8\,\text{m}\,\text{s}^{-2}$.

1 Calculate the force required to accelerate a jumbo jet of mass $4.0 \times 10^5\,\text{kg}$ to $80\,\text{m}\,\text{s}^{-1}$ in $15\,\text{s}$ at take off. (3)

2 A cyclist of mass $70\,\text{kg}$ travelling at $4.5\,\text{m}\,\text{s}^{-1}$ crashes into a brick wall and comes to rest in $0.20\,\text{s}$. Calculate:
   (a) The deceleration (negative acceleration) of the cyclist. (2)
   (b) The force exerted on the cyclist by the wall. (2)

3 A girl of mass $55\,\text{kg}$ rides in a lift.
   (a) Calculate her weight. (1)
   (b) State the force exerted on her by the lift when it is at rest. (1)
   (c) Calculate the force exerted on her by the lift when it is accelerating upwards at $1.4\,\text{m}\,\text{s}^{-2}$. (2)
   (d) Draw a free-body force diagram for the girl in case (**c**). (2)

4 A student drops a stone down a well and records that it takes $2.5\,\text{s}$ to reach the water surface. Calculate:
   (a) The depth of the water surface below ground. (2)
   (b) The speed of the stone as it hits the water surface. (2)

5 In which of these is there **(a)** zero resultant force, and **(b)** zero resultant moment? (3)

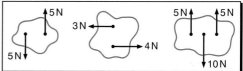

6 An athlete supports a weight of $20\,\text{N}$ at arm's length, $0.70\,\text{m}$ from his shoulder joint. State the moment of the weight twisting his shoulder. (1)

7 A metre rule is a uniform rod.
   (a) Where is the centre of gravity? (2)
   (b) The rule is pivoted at the $20\,\text{cm}$ mark and balances when a weight of $1.8\,\text{N}$ is hung from the $0\,\text{cm}$ mark. Calculate the weight of the rule. (2)

The answers are on pages 104–105.

Work

When **work** is done **energy is transformed**. Mechanical work is done when a force moves in the direction of the force.

- Work done = force × displacement, $W = fd$, in the direction of the force.
- **Force** and **displacement** are both **vectors** but **work** is a **scalar**. Thus energy is a scalar.
- When the force is at an angle θ to the direction of movement the work done is $Fd \cos \theta$.
- The component of force F parallel to the direction of movement is $F \cos \theta$ and this does work when it moves in the direction of d.
- The component $F \sin \theta$ is perpendicular to the displacement and does no work.

The unit of work and of energy is the joule (J).

- 1 joule of work is done when a force of 1 newton moves 1 metre in the direction of the force.

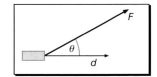

Examples

- When you raise a weight W through a vertical height Δh you do work $W\Delta h$ because the vectors W and Δh are parallel.
- When you carry a weight W through a horizontal displacement, no work is done because the weight and displacement vectors are perpendicular to each other.
- When you lean on a wall, exerting a force F on the wall, no work is done unless the wall moves!

Gravitational potential energy

When work is done in raising a mass m, that work in transformed into an increase in the gravitational potential energy ΔE_P of the mass. The work done in moving the mass a vertical distance Δh is $mg\Delta h$ since weight $W = mg$.

- the change in gravitational potential energy $\Delta E_P = mg\Delta h$.

Kinetic energy

When work is done accelerating a body, that work is transformed into the kinetic energy of the body.

Suppose a body is accelerated with acceleration a to reach a velocity v in distance s. Combine the equation of motion $v^2 = 2as$ with $F = ma$.

Substituting $a = \frac{F}{m}$ in $v^2 = 2as$ we get $v^2 = \frac{2Fs}{m}$.

Now Fs is the work done accelerating the body and Fs equals $\frac{1}{2}mv^2$.

Thus $\frac{1}{2}mv^2$ is the kinetic energy of the body.

● The kinetic energy E_k of a moving body $= \frac{1}{2}mv^2$.

Worked example

Find the velocity of a stone of mass 3.0 kg, which has fallen 25 m under gravity by equating the drop in gravitational potential energy to the gain in kinetic energy.

Formula required is $\Delta E_P = mg\Delta h$.

Data given are $m = 3.0$ kg, $g = 9.8$ N kg^{-1}, $\Delta h = 25$ m.

Drop in $E_P = (3.0\,\text{kg})(9.8\,\text{N kg}^{-1})(25\,\text{m}) = 735\,\text{J} = $ gain in E_k.

Substituting, $\frac{1}{2}(3.0\,\text{kg})v^2 = 735\,\text{J}$ gives $v = 22\,\text{m s}^{-1}$.

The velocity of the stone is $22\,\text{m s}^{-1}$.

Power

Power is the rate of doing work or the **rate of transfer of energy**.
● Power = energy transferred divided by time taken. $P = \frac{E}{t}$
● Power = work done divided by time taken.
● Power = force × displacement/time (since work = force × displacement)
    = force × velocity. $P = Fv$
● The unit of power is the watt. 1 watt = 1 joule per second.
    $(W = J s^{-1})$
● Electrical power = p.d. × current (see page 38)

Remember the relationship between energy and power. For a given fuel supply (energy input), an engine run at twice the power will burn the fuel at twice the rate and run out of fuel in half the time.

Worked example

A car engine develops 65 kW at a steady speed of $25\,\mathrm{m\,s^{-1}}$. Calculate the force at the wheels driving the car.

The formula required is $P = Fv$, rearranged to give $F = \dfrac{P}{v}$.

Substituting gives $F = \dfrac{65\,000\,\mathrm{W}}{25\,\mathrm{m\,s^{-1}}} = 2600\,\mathrm{N}$.

The force propelling the car is $2600\,\mathrm{N}$.

The conservation of energy

The conservation of energy is one of the most fundamental laws of physics. It states that **energy may be transformed from one form to another, but not created nor destroyed**.

You will be familiar with many energy **transformations** or **transfer methods**. For example:
- electrical to mechanical: an electric motor
- mechanical to electrical: an alternator in a power station
- gravitational to kinetic: water flowing over a waterfall
- chemical to electrical and thermal: a coal-fired power station
- chemical to mechanical and thermal: a petrol engine
- electrical to light: an LED.

For any such transformation we can draw up an energy balance equation:

total energy of system at start = total energy of system at finish.

To use this equation we need formulae to calculate the energy associated with each transformation or transer method. Always calculate energies in joules.

Here is a list of the energy formulae you need to know about.
- kinetic energy $E_k = \frac{1}{2}mv^2$ (page 23)
- gravitational potential energy $\Delta E_P = mg\,\Delta h$ (page 22)
- stored energy in spring $E = \frac{1}{2}Fd$ (page 32)
- electrical energy $E = Vq = VIt$ (page 38)
- thermal energy $\Delta E_{Th} = mc\Delta\theta$ (page 93)
- thermal energy $\Delta E_{Th} = mL$ (page 93).

Check yourself

Work and energy

1 Calculate the work done when a workman
   (a) lifts a barrel of mass 25 kg from the ground onto a truck 1.2 m above the ground. (2)
   (b) rolls the same barrel up a plank onto the same truck. (2)

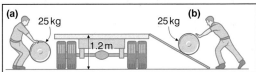

2 **(a)** Calculate the kinetic energy of a skateboarder of mass 65 kg moving at 5.5 m s^{-1}. (1)
   (b) The skateboard park contains a ramp. Neglecting friction, calculate the vertical height the skateboarder will rise up the ramp before coming to rest. (2)

3 Calculate the speed of an alpha particle with kinetic energy of 8.0 × 10^{-13} J (mass of alpha particle = 6.4 × 10^{-27} kg). (3)

4 150 kg of water flows through a water turbine each second. The water falls 3.5 m as it passes through the turbine. Calculate the gravitational potential energy transformed each second and, hence, the maximum power input to the turbine. (3)

5 A student of mass 70 kg runs up the stairs between two floors 4.0 m apart in 5.0 s. Calculate the power he develops. (3)

6 Use the equation $P = Fv$ to show that a car with an engine providing constant power has an acceleration that falls with increasing speed as the car nears its top speed. (3)

7 **(a)** A skier, mass 75 kg, skis downhill for 2000 m through a vertical height of 200 m. Calculate the fall in his gravitational potential energy. (2)
   (b) He is subject to a frictional force of 24 N. Calculate the energy transformed in doing work against friction. (2)
   (c) Assuming that the rest of the gravitational potential energy is transformed into kinetic energy, calculate his speed at the bottom of the hill. (2)

The answers are on pages 105–106.

Friction and resistive forces

We are all familiar with frictional forces, which **oppose motion**. For example, to slide a book across the table you have to do work against frictional forces. This work is transformed into thermal energy and usually results in an increase in the internal energy of the surroundings.

- Frictional forces always act in the opposite direction to motion.
- The force necessary to start two solids sliding is greater than that needed to keep them sliding.
- Frictional forces in fluids are called **viscous** forces, and the fluid has the property of **viscosity**. Viscosity decreases with increase in temperature.
- Frictional forces in air (**drag** or **air resistance** forces) increase as the velocity of the object increases.

Force pairs

When a block slides along a table the frictional force (F_1) exerted on the block by the table is equal in magnitude and opposite in direction to the force (F_2) exerted on the table by the block.

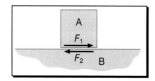

This is a statement of Newton's third law: **action and reaction forces are equal and opposite and act on different bodies.**

Worked example

The frictional force on a train moving at 4.5 m s^{-1} is 25 kN. Calculate (a) the work done against friction to move the train 4.5 m, and (b) the power needed to overcome friction at 4.5 m s^{-1}.

(a) Formula required: work done $= Fd$.

Data given: $F = 25\,\text{kN} = 25\,000\,\text{N}$, $d = 4.5\,\text{m}$

Substituting: work done $= (25\,000\,\text{N})(4.5\,\text{m}) = 113\,\text{kJ}$.

(b) Power $=$ work done per second $=$ force $\times$ velocity.

Substituting gives power $= (25\,000\,\text{N})(4.5\,\text{m s}^{-1}) = 113\,\text{kW}$. (*Note that the power necessary to overcome friction rises as the speed rises.*)

Air resistance and terminal velocity

(Required for AQA(B), Edexcel(SH), OCR(A & B) and WJEC only)

When a body moves through a fluid (a liquid or a gas) it experiences a **drag** or **resistive force**, which increases as the velocity of the body increases. For many situations the **drag force is proportional to (velocity)^2**, $F \propto v^2$.

When the drag force is equal and opposite to the force accelerating the body the resultant force is zero. The body then moves at a constant velocity called the **terminal velocity**.

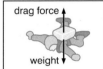

Consider a skydiver jumping out of an aircraft. Initially his weight is greater than the drag force so he accelerates downwards. As his velocity increases the drag force increases until it is equal to and opposite to his weight. From this point on he falls at his terminal velocity in air. Opening his parachute makes a large increase in the drag force at a given velocity, so he slows down until the drag force is once again equal and opposite to his weight at a much lower velocity.

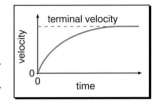

Worked example

A car manufacturer quotes the 'drag coefficient', C_D, for a motor car. The drag force is given by $F = C_D v^2$. For a particular car $F = (1.1\,\mathrm{kg\,m^{-1}})v^2$. The car engine produces a motive force of 2000 N at its top speed. Calculate the top speed, or terminal velocity, of the car.

At the top speed drag force = motive force.

Substituting in the formula for drag force, $v^2 = \dfrac{2000\,\mathrm{N}}{1.1\,\mathrm{kg\,m^{-1}}}$

Working out v gives a top speed for the car of $47\,\mathrm{m\,s^{-1}}$.

Forces on aircraft
(Required for AQA(B) only)

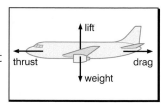

When the aircraft is in level flight at constant velocity:
- The lift force is equal and opposite to the weight.
- The thrust force is equal and opposite to the drag force.

Forces on cars
(Required for OCR(A) and other general questions on motion.)

The engine provides the **motive force** to propel the car. It turns the wheels to exert a **backwards** force on the road. Provided there is sufficient friction between the tyre and the road surface, the road exerts an equal forwards force on the car to propel it forwards. (These forces form a pair – see page 26.)

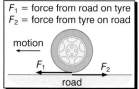

- When the motive force is greater than the drag force the car accelerates.
- When the motive force equals the drag force the car maintains constant velocity.
- When the motive force is replaced by a braking force the car slows down under the action of the braking force and the drag force.

Cars rely on friction to stay on the road. For safety there should be no slip between the tyre and the road. When the tyre skids or slides the frictional force is much less and the car is very difficult to control. To transmit the large forces involved, modern tyres are made wide, to give a large area in contact with the road, and with a grooved tread to disperse water when the road is wet.

Worked example
At a speed of 25 m s^{-1} the drag force on a car of mass 800 kg is 500 N. The engine of the car produces a motive force of 1200 N. Calculate the acceleration of the car.

The resultant force accelerating the car = 1200 N − 500 N = 700 N
The acceleration of the car $a = \dfrac{F}{m} = \dfrac{700N}{800kg} = 0.88\,\text{m s}^{-2}$.

Stopping distances for cars

The distance travelled by a car when stopping is the sum of the distance travelled while the driver reacts to the emergency and the distance travelled whilst braking:

stopping distance = thinking distance + braking distance.

Worked example

A driver sees a child run into the road. The car is moving at 17 m s^{-1} and the brakes can decelerate the car at 3.5 m s^{-2}. If the driver takes 0.40 s to apply the brakes, calculate how far the car travels before it stops.

Thinking distance = ut = $(17 \, \text{m s}^{-1})(0.40 \, \text{s})$ = 6.8 m.

Braking distance is found from the formula $v^2 = u^2 + 2 \, as$.

Data given are $u = 17 \, \text{m s}^{-1}$, $v = 0$, $a = -3.5 \, \text{m s}^{-2}$.

Substituting gives $0 = (17 \, \text{m s}^{-1})^2 - 2s(3.5 \, \text{m s}^{-2})$. Rearranging gives $s = 41.3 \, \text{m}$

Braking distance = 41.3 m. Total stopping distance = 48 m.

When a car is involved in a crash it comes to rest very quickly. The very large deceleration means that the driver and passengers are subjected to very large forces. All the kinetic energy of the car must be dissipated as it comes to rest. To help this happen, modern car bodies have crumple zones and the people inside them are protected by seat belts and air bags.

- **Crumple zones.** Much of the kinetic energy of the car can be dissipated in crushing the sections of the car outside the cage round the passengers. When a car runs head-on into a wall, although the front of the car stops, the people continue to move forward as the front crumples. They move further in stopping than the car does and so are subject to smaller decelerations – and thus smaller forces.
- **Seat belts and air bags** restrain the passengers and dissipate their kinetic energy. Seat belts dissipate energy by progressively tearing; when a person hits the air bag it deflates progressively, dissipating energy. Both systems allow the passenger to move forward a bit and so increase the stopping distance – and hence reduce the force on them.

Momentum and the conservation of momentum

(Required for AQA(A), CCEA and Edexcel only)

- Momentum (p) is defined as **mass** $\times$ **velocity**, $p = mv$.
- Momentum is a **vector** quantity.
- The unit for momentum is **N s** or **kg m s^{-1}**.
- Force = rate of change of momentum, $F = \frac{\Delta(mv)}{\Delta t}$.
- Change in momentum (= force $\times$ time) is called impulse.

The conservation of momentum

Momentum is a useful concept, particularly when dealing with collisions. In a system on which no external forces act, momentum is conserved.

In the absence of friction, the total momentum of two bodies colliding is the same after the collision as it was before.

This is Newton's third law of motion. The forces between the two bodies form a pair, which are equal and opposite to each other.

Elastic and inelastic collisions

The kinetic energy of the bodies may change in a collision.

- In an **elastic** collision, kinetic energy is **conserved**. For example, two helium atoms in helium gas collide elastically.
- In an **inelastic** collision, kinetic energy is **not conserved**. The total kinetic energy before the collision is greater than the total kinetic energy afterwards. For example, a car colliding with a wall is an inelastic collision.

Worked example

A ball of mass 1 kg moving at 3 m s^{-1} collides with a stationary ball of mass 4 kg and sticks to it. Calculate the velocity of the balls after the collision.

Apply conservation of momentum.

Total momentum before = $(1\,\text{kg})(3\,\text{m s}^{-1}) + 0 = 3\,\text{N s}$.

Total momentum after = $(5\,\text{kg})v$

Substituting gives $v = \frac{3\,\text{Ns}}{5\,\text{kg}} = 0.60\,\text{m s}^{-1}$.

Velocity of balls after the collision = $0.60\,\text{m s}^{-1}$.

Forces and motion

1 A cyclist experiences a constant frictional force of 15 N.
   (a) Calculate the work done against friction in cycling 1.0 km. (2)
   (b) Calculate the power required to overcome friction in cycling that kilometre in 1 minute. (2)

2 Identify the force pair when you are supporting this book on your hand. (2)

3 Explain why the terminal velocity of a person in free fall is much greater than their terminal velocity when attached to an open parachute. (3)

4 The drag force on a falling object of mass 2.0 kg is 10 N at a speed of $30\,\mathrm{m\,s^{-1}}$ and 20 N at a speed of $60\,\mathrm{m\,s^{-1}}$. Calculate the acceleration of the object at $30\,\mathrm{m\,s^{-1}}$ and $60\,\mathrm{m\,s^{-1}}$ and describe its motion. (4)

5 Calculate the force on a car passenger of mass 65 kg, necessary to stop her from a speed of $30\,\mathrm{m\,s^{-1}}$ in a distance of 1.6 m. (4)

6 Show that the braking distance for a car is multiplied by four when the initial speed of the car is doubled. (3)

7 Calculate the momentum of an electron moving at a velocity of $2.0 \times 10^7\,\mathrm{m\,s^{-1}}$. (2)

8 A bullet of mass 10 g travelling at $300\,\mathrm{m\,s^{-1}}$ embeds itself in a box of sand of mass 8.0 kg. Calculate:
   (a) The momentum of the bullet before colliding with the box. (2)
   (b) The velocity of the box + bullet after the collision. (2)

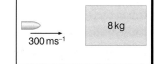

8 kg

$300\ \mathrm{ms^{-1}}$

9 A collision between two identical steel balls, A and B, may be regarded as elastic. Ball A collides with a stationary ball B. A stops and B moves off. Show that the conservation of momentum and of kinetic energy can be satisfied when the velocity of B after the collision is the same as that of A before the collision. (5)

The answers are on pages 106–107.

(**Not** required for AQA(B) or CCEA)

When **tension** forces are applied to the ends of a bar of material the bar **stretches**. When **compression** forces are applied, the bar is **compressed**. Both tension and compression can result in the failure of the bar.

Stretching springs

As the **load** (the stretching force) is increased, the **extension** (the increase in length of the spring) increases. For some materials the deformation is **elastic**.

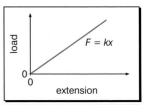

- **Elastic** materials obey **Hooke's law**, $F = kx$.
- For elastic materials, the extension x is proportional to the force F.
- The k in $F = kx$ is called the **spring constant** or **stiffness** and has units $N\,m^{-1}$.
- The work done in stretching a spring equals the area between the load/extension graph and the extension axis. For elastic materials, work done $= \frac{1}{2}Fx = \frac{1}{2}kx^2$.
- All the work done in stretching a material elastically is recovered when the material is unloaded.

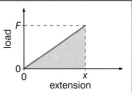

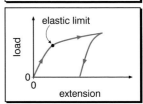

Plastic deformation occurs when a material does not return to its original shape when the load is removed. Materials that show plastic deformation:

- are permanently deformed when unloaded
- do not obey Hooke's law ($F = kx$) in the plastic region
- do not return the work done in stretching when they are unloaded.

Many materials have an elastic region, where Hooke's law is obeyed, for small loads. When the load exceeds the **elastic limit** the material deforms plastically.

Three quantities enable us to make calculations on specimens of a material of different shapes and sizes:

- **Stress = load force/area of cross-section**, stress $\sigma = \frac{F}{A}$, is measured in pascals, **Pa**. Although the unit is the same as that for pressure, stress occurs within the solid material whilst pressure is exerted on the surface of the solid material.

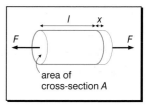

- **Strain = extension/original length**, strain $\varepsilon = \frac{x}{l}$. This is a ratio and has no units.
- **The Young modulus** of elasticity = **stress/strain**, $E = \frac{Fl}{Ax}$. The Young modulus of a material is a characteristic property of that material. It is measured in $N\,m^{-2}$ or **Pa**. It is measured near the origin of the stress/strain graph where the deformation is usually elastic. Stress/strain graphs are the same for all samples of a given material. Load/extension graphs relate to a particular sample.
- The **ultimate tensile stress** (UTS) is the stress that causes failure of the material.
- The stored energy per unit volume = $\frac{1}{2}$ **stress** × **strain** for an elastic deformation. It is the area between the graph and the strain axis on a stress/strain graph.
- A **stiff** material is one with a large Young modulus.

Different types of materials

Brittle materials, such as glass, obey Hooke's law up to the point of failure. They show no plastic deformation. They fail suddenly by crack propagation. They shatter when broken because the stored energy is transformed into kinetic energy of the fragments.

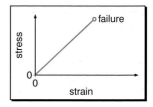

Ductile materials, such as copper, obey Hooke's law at very small stresses but show considerable plastic deformation beyond the elastic limit. Copper can be drawn out into wires (it is ductile) and beaten into sheets (it is **malleable**).

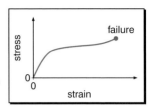

Polymeric materials, such as rubber or polyethene, do not obey Hooke's law – the stress/strain or load/extension graph has no straight line section. Many polymeric materials exhibit **hysteresis** – the graph for unloading the material does not coincide with that for loading the material.

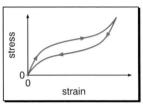

Tough materials have a large stored energy per unit volume. A large amount of work must be done by plastic deformation to make them fail. For example, to break a piece of mild steel a lot of work has to be done in deforming it before it fails. It is difficult to propagate cracks in tough materials.

Fatigue failure occurs in materials that fail after repeated application of a stress well below the ultimate tensile stress. For example, aluminium bolts fail when subjected to a lot of vibration.

Creep occurs when a material continues to extend under a constant applied load – such as the handles of a polyethene shopping bag that stretch when carrying the shopping home!

Check yourself

Properties of solids

1 A spring with a spring constant of $250\,\mathrm{N\,m^{-1}}$ supports a load of
15 N. Calculate
(a) The extension of the spring. (2)
(b) The stored energy in the spring. (2)

2 A student has a supply of springs, each with a spring constant of
$25\,\mathrm{N\,m^{-1}}$. Find the spring constant for assemblies of
(a) Two springs in series. (2)
(b) Two springs in parallel. (2)

3 The rear suspension of a motor car compresses by 5.0 cm when
a load of 60 kg is placed over the rear axle. Two springs support
the load. Calculate the spring constant of each spring. (4)

4 The load/extension graph shows the result
of a laboratory experiment on a spring.
(a) Calculate the spring constant for
small loads. (2)
(b) Describe what you think happened as
the load was increased and later
reduced. (3)

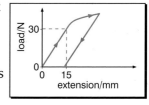

5 A steel cable of cross-sectional area $2.4 \times 10^{-6}\,\mathrm{m^2}$ supports a
load of 150 N.
(a) Calculate the stress in the cable. (2)
(b) The original length of the cable is 5.0 m and the Young
modulus of the steel used is $2.1 \times 10^{10}\,\mathrm{Pa}$. Calculate the
extension of the cable. (3)

6 Give one advantage and one disadvantage of using ductile
materials in the construction of motor cars. (4)

7 The graph shows the stress/strain graph
for a metal. Work out as many properties
of the metal as you can from the graph.
(6)

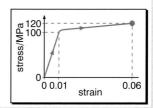

The answers are on pages 107–108.

Electric current

An electric current is flow of electric charge around a circuit.

- The unit of electric current is the **ampere** (A), symbol I.
- The unit of electric charge is the **coulomb** (C), symbol q.
- One coulomb is the charge that flows when a current of one ampere is set up for one second:

$$I = \frac{\Delta q}{\Delta t} \text{ or } \Delta q = I\Delta t$$

The ampere is one of the base SI units.

Electric charge is usually calculated from current × time using the equation above.

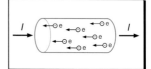

In different materials the charge carriers that form the current are different.

- In **solids**, the moving charges are always **electrons**.
- In **liquid metals**, the moving charges are also **electrons**.
- In other **liquids**, the charges are carried by positive and negative **ions**. For example, in water the ions are H_3O^+ and OH^-.
- In **gases**, the charges are carried by positive and negative **ions**.

Whatever the sign and nature of the charge carriers in any given material, all electrical calculations assume that charge flows from the positive terminal to the negative terminal of the cell or power supply.

Although the electron has a negative charge and moves from negative to positive around a circuit (remember, unlike charges attract), arrows on circuit diagrams are always drawn on the assumption that current is from positive to negative!

The conservation of charge

Electrons cannot be created or destroyed in circuits so the **total charge in a circuit is constant**. At any point in a circuit the current entering that point must equal the current leaving it. This is known as **Kirchhoff's first law**.

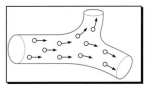

At any junction in a circuit, the current entering the junction must equal the current leaving the junction.

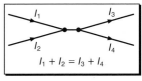

$$I_1 + I_2 = I_3 + I_4$$

In a wire carrying a current the current is the quantity of charge passing per second. It depends on the number of charge carriers per m^3 (n), the area of cross-section of the wire (A) and the drift velocity at which the carriers move along the wire (v). The charge on each carrier is e. In a solid e is the charge on an electron. The number of electrons passing any point in one second is nAv, having a charge $nAve$. Thus the current is given by:

$$I = nAve$$

For copper, n is approximately $10^{29}\,m^{-3}$, leading to a drift velocity of the order $1\,mm\,s^{-1}$ for currents of a few amperes.

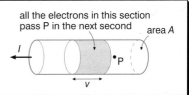

all the electrons in this section pass P in the next second   area A

Cells and electrical energy

To set up a current in a circuit a source of energy is required. This can be a cell or battery, or a mains power supply.

The energy (in joules) supplied by the cell to each coulomb is called the **electromotive force** (emf) of the cell or power supply. It is measured in **volts** (V), and is usually given the symbol E.

When the current transfers this energy around a circuit to a component such as a lamp the energy transformed in the lamp by each coulomb passing through is the **potential difference** (p.d. or voltage) across the lamp.

- p.d. is also measured in volts.
- The usual symbol for p.d. is V.
- One volt = one joule per coulomb:

$$V = \frac{W}{q} \text{ or } W = qV$$

Electrical power

Power, P, is energy divided by time: $P = \frac{W}{t}$. Since $W = qV$ and $I = \frac{q}{t}$:

$$P = VI$$

Another rearrangement gives energy $W = VIt$ since $q = It$.

Worked example

A car headlamp is rated at 60 W and runs from a 12 V battery. Calculate the current in the lamp and the charge flowing through the lamp in 1 minute.

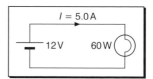

Formula required: $P = VI$.

Data given: $P = 60\,W$, $V = 12\,V$.

Substituting: $I = \frac{P}{V} = \frac{60\ W}{12\ V} = 5\ A$.

$5\,A = 5\,C\,s^{-1}$, so in $60\,s$ charge flowing is $(5\,C\,s^{-1})(60\,s) = 300\,C$.

The current is 5 A and 300 C flow through the lamp in 60 s.

Electrical resistance and conductance

Any component with a potential difference across it is said to have **resistance**.

● Resistance is p.d./current.
● The unit of resistance is the **ohm** (Ω) and the symbol is R.
● A resistor is a component with the property of resistance.
● 1/resistance is called conductance and is defined as current/p.d.
● The unit of conductance is the **siemens** (S = Ω^{-1}) and the symbol is G.
● A high resistance equates to a low conductance.

For some resistors, such as metals at constant temperature, the resistance is constant and the current is proportional to the p.d. across the resistor. Such a resistor obeys **Ohm's Law**:

$$V = IR$$

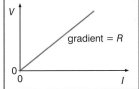

A graph of V against I is a straight line with a gradient of R.

Combining the power relationship $P = VI$ with Ohm's law gives two further useful relationships:

$$P = I^2R = \frac{V^2}{R}$$

Now you can set up a complete circuit with the appropriate circuit symbols (see page 119). Remember, the ammeter is in series with the component (it measures the current *in* the component) and the voltmeter is in parallel with the component (it measures the p.d. *across* the component).

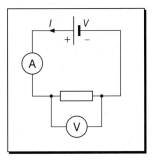

Combinations of resistors in series

When resistors are connected in series:

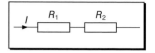

- there is the **same current** in each resistor
- the supply **p.d.** is **shared** across the set of resistors
- the total potential difference across the set of resistors is the sum of the potential differences across each resistor in the set.

Using $V = IR$ we can show that the total resistance R_T of resistances R_1, R_2, ... in series is given by

$$R_T = R_1 + R_2 + ...$$

Combinations of resistors in parallel

When resistors are connected in parallel:

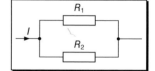

- there is the **same p.d.** across each resistor
- the **current** from the supply is **shared** between the resistors
- the total current is the sum of the currents in each of the set of resistors.

Using $V = IR$ we can show that the total resistance R_T of resistances R_1, R_2, ... in parallel is given by:

$$\frac{1}{R_T} = \frac{1}{R_1} + \frac{1}{R_2} + ...$$

(In terms of conductance: $G_T = G_1 + G_2 + ...$)

Worked example

Two resistors, each of resistance 20 Ω, are connected (a) in series, and (b) in parallel. Calculate the total resistance of each combination.

(a) Using the formula for resistors in series gives:
$R_T = (20\,\Omega) + (20\,\Omega) = 40\,\Omega$.
(Two equal resistors of value R in series have a total resistance of $2R$.)

(b) Using the formula for resistors in parallel gives:
$\frac{1}{R_T} = \frac{1}{20\,\Omega} + \frac{1}{20\,\Omega} = \frac{1}{10\,\Omega}$. $R_T = 10\,\Omega$.

(Two equal resistors of value R in parallel have a total resistance of $\frac{1}{2} R$.)

Electricity basics

1 How many coulombs of charge pass through a wire carrying a current of 3.0 A in 12 s? (2)

2 The current in a wire increases linearly from zero to 4.0 A in 20 s. Plot a sketch graph of current against time and use it to find the total charge transferred. (3)

3 **(a)** In which direction will electrons move in a circuit? (1)
   (b) How is this related to the conventional current? (1)

4 The charge on one electron is 1.6×10^{-19} C. How many electrons pass per second through a wire carrying 1 A? (3)

5 A 1.5 V dry cell can circulate 2500 C around a circuit before it is exhausted. How much energy has it delivered? (2)

6 A 'Walkman' is powered by a 3.0 V battery and draws 40 mA when operating. Calculate the power provided by the battery and the energy transferred in listening to a tape for 1 hour. (3)

7 Calculate the current in a 10 Ω resistor when it is connected to a 6.0 V battery. (2)

8 Calculate the current in a 24 Ω resistor at which 15 W is dissipated in the resistor. (2)

9 The graph shows a V–I graph for a resistor which obeys Ohm's Law. Find the resistance of the resistor from the graph. (1)

10 A student is given resistors labelled 4 Ω and 5 Ω. Calculate the resistance of the pair.
   (a) In series. (2)
   (b) In parallel. (2)

The answers are on pages 108–109.

Resistance and resistivity (conductance and conductivity)

Resistors come in many shapes and sizes. The resistance of a particular resistor depends on the material it is made of, its length (l) and its cross-sectional area (A) – assuming it is of uniform cross-section.

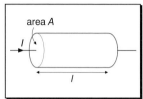

- The constant for the material is called its **resistivity** and is denoted by ρ. The unit of resistivity is $\Omega\,m$.
- For conductance the constant is conductivity, denoted by σ. The unit of conductivity is $S\,m^{-1}$.
- Resistance of a wire is **proportional** to **length** (double the length, double the resistance).
- Resistance of a wire is **inversely proportional** to the area of **cross-section** (double the area, halve the resistance).

The formulae relating these quantities are:

$R = \dfrac{\rho l}{A}$ for resistance, $G = \dfrac{\sigma A}{l}$ for conductance.

The resistivities of materials vary enormously – from $10^{-8}\,\Omega\,m$ for good conductors like copper to $10^{12}\,\Omega\,m$ for good insulators like Perspex.

Resistivity varies with temperature:
- The resistance of metals increases with increase of temperature.
- The resistance of semiconductors such as the negative temperature coefficient (NTC) thermistor decreases with increase of temperature.

The graphs show the variation of resistance with temperature for a metal wire and for an NTC thermistor between 0 °C and 100 °C.

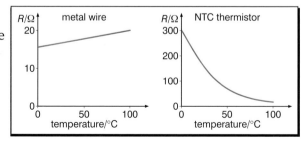

Ohmic and non-ohmic resistors

Ohm's law ($V = IR$) can only be applied to circuit calculations where the resistance of the resistor is constant. Such resistors are called **ohmic resistors**. Metals kept at constant temperature are the most common type of ohmic resistor.

The *V–I* graph of an ohmic resistor is a straight line through the origin. A resistor that gives a non-linear *V–I* graph is said to be **non-ohmic**.

A **filament lamp** is non-ohmic because the filament gets hotter as the power dissipated in it is increased. The resistance of the filament increases with increase in temperature and is much greater when the filament is white hot than when cold.

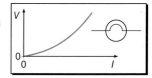

An **NTC thermistor** is non-ohmic because, when power is dissipated in it and it gets hot, the resistance falls and the current increases rapidly.

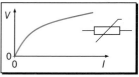

A **semiconductor diode** is non-ohmic because in the forward direction it passes a current when the p.d. across it is greater than 0.7 V but in the reverse direction the current is almost zero.

A **light-emitting diode** (**LED**) has a graph similar to that of the semiconductor diode. In the forward direction it conducts and emits light when the p.d. across it is greater than 1.6 V.

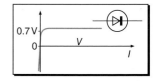

A **light-dependent resistor** (**LDR**) is non-ohmic because its resistance changes with light intensity. It is commonly used as a light sensor.

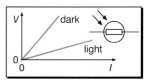

The mechanism of conduction in solids

● In **metals** there are very many 'free' electrons, which are not bound to particular ions in the lattice and can drift through the lattice when a p.d. is applied. Resistance arises as these 'free' electrons collide with ions in the lattice and lose the energy gained from the p.d. At higher temperatures the resistance is greater because the amplitude of the lattice vibrations increases and the electrons collide with the lattice more often.

● In **semiconductors** there are few 'free' electrons and so the resistance is much higher. When the temperature rises many more electrons are 'freed' and the resistance falls. In a LDR the increase in the number of 'free' electrons comes from the light energy falling on the LDR.

● In **insulators** there are no 'free' electrons to carry the current.

Worked example

A mains lamp is rated at 60 W, 240 V. The lamp filament is made of tungsten wire of length 0.40 m. Tungsten wire has a resistivity of $5.9 \times 10^{-6}\,\Omega\,m$ at its normal operating temperature. Calculate the diameter of the wire.

First work out the resistance of the lamp filament.

Formula required: $P = \dfrac{V^2}{R}$, rearranged to give $R = \dfrac{V^2}{P}$. Data given:

$P = 60\,W, V = 240\,V.$

Substituting gives $R = \dfrac{240\,V^2}{60\,W} = 960\ \Omega.$

Now work out the area of cross-section A.

Rearrange the formula $R = \dfrac{\rho l}{A}$ to get $A = \dfrac{\rho l}{R}$ and substitute:

$A = \dfrac{(5.9 \times 10^{-6}\,\Omega\,m)\ 0.40\,m}{960\,\Omega} = 2.5 \times 10^{-9}\,m^2$

The wire is circular so the area $A = \pi r^2 = \dfrac{\pi(\text{diameter})^2}{4}$.

Rearranging this formula gives the diameter $= \sqrt{\dfrac{4A}{\pi}} = 5.6 \times 10^{-5}\,m.$

Check yourself

Electrical resistance

1 A student wishes to make a 2.5 Ω resistor from nichrome wire of area of cross-section $6.5 \times 10^{-8}\,m^2$. Nichrome has a resistivity of $1.1 \times 10^{-6}\,\Omega\,m$. Calculate the length of wire required. (2)

2 A lump of conducting putty is rolled into a cylinder and contact is made to each end. The resistance is 15 Ω. The same lump of putty is now rolled out to make a cylinder three times as long. Calculate the resistance of the new cylinder of putty. (3)

3 A coil of fine copper wire has a resistance of 1.50 Ω at 20 °C. A p.d. of 1.08 V is connected across it. Calculate the current in the copper wire. The coil is now placed in boiling water at 100 °C and its resistance is 1.98 Ω. Has the current in the coil increased or decreased? By how much? (3)

4 The graph shows the $\frac{V}{I}$ plot characteristic for a filament lamp. Explain how to find the resistance of the lamp at any particular current in the lamp. Describe how the resistance of the lamp changes as the current increases. (2)

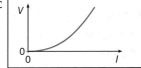

5 The circuit shown contains two semiconductor diodes. Explain which lamps will light when the switch is closed. (2)

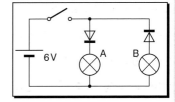

6 The circuit for a simple light meter is shown. Explain which end of the meter scale would correspond to darkness and which to sunlight. (2)

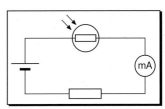

The answers are on page 109.

Look at this circuit.

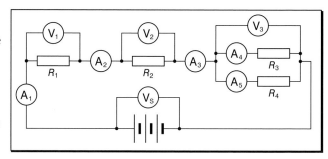

- Ammeters A_1, A_2 and A_3 will all register the same current because resistors R_1 and R_2 are in series. (This is conservation of charge: no charge gets 'lost' in the resistor.)
- The current registered by A_3 is the sum of the currents registered by A_4 and A_5 because the current is shared between resistors R_3 and R_4 in parallel. (This is also conservation of charge: charge does not get 'lost' at the junction.)
- Voltmeters V_1, V_2 and V_3 register the p.d. across R_1, R_2 and the parallel combination of R_3 and R_4. The sum $V_1 + V_2 + V_3$ must equal the terminal p.d. V_S across the battery or supply. (This is the conservation of energy: each coulomb receives V_S joules from the battery to dissipate around the circuit.)

The internal resistance of a supply

Any cell or power supply has an **internal resistance** associated with it. This includes the resistance of the chemicals in the cell and the resistance of any wires within the battery or power supply.

Any charge that goes round a complete circuit must go through the internal resistance of the cell. Energy is required to take the charge through the cell, so the energy per unit charge available to do work outside the cell is less than the emf (page 38) of the cell. Some of the emf is 'used up' within the cell.

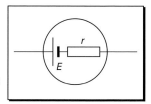

- The terminal p.d. (V_S) across a cell supplying a load resistor is less than the emf (E) of the cell. The equation is:

$$V_S = E - Ir$$

where I is the current in the cell and r is the internal resistance of the cell.

- In circuit diagrams a cell with internal resistance is shown by an ideal cell of emf E in series with a resistance r.

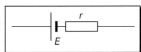

- The energy dissipated by the internal resistance heats the cell.
- The cell delivers maximum power when the load resistance equals the internal resistance.
- The cell works most efficiently when the load resistance is much greater than the internal resistance.

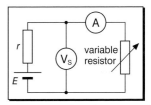

- To find the emf and internal resistance of a cell, measure the terminal p.d. V_S and current I for a range of values of load resistance. Plot a graph of V_S against I. The gradient of the graph is $-r$ and the intercept on the V_S axis is E.

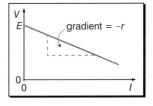

Worked example

A cell of emf 1.50 V and internal resistance 0.75 Ω is connected to a lamp of resistance 2.5 Ω. Calculate the current in, and p.d. across, the lamp.

The p.d. across the lamp is the terminal p.d. V_S of the cell.

Equations required: $V_S = IR$ for the lamp, $V_S = E - Ir$ for the cell.

Data given: $E = 1.50\,V$, $r = 0.75\,\Omega$, $R = 2.5\,\Omega$.

Substituting gives $I = \dfrac{E}{R + r} = \dfrac{1.50V}{2.5\Omega + 0.75\Omega} = 0.46\,A.$

$$V_S = IR = (0.46\,A)(2.5\,\Omega) = 1.15\,V.$$

The current in the lamp is 0.46 A and the terminal p.d. across the lamp is 1.15 V.

Kirchhoff's second law

Kirchhoff's second law is a statement of the conservation of energy in electrical circuits. As each coulomb goes round the circuit it passes:

- the battery and picks up 12 J
- the internal resistance of the battery and dissipates 1 J
- the lamp and dissipates 8 J
- the resistor and dissipates 3 J

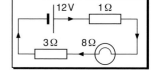

and then returns to the battery.

Put in mathematical form the law states that, in any circuit, the sum of IR for each of the components going round the circuit equals the emf E in the circuit:

$E = \sum(IR)$ around the circuit.

The potential divider

The potential divider enables us to produce fractions of a given supply p.d. or, with a variable resistor or potentiometer, to produce a variable p.d. It consists of two resistors connected in series across the supply.

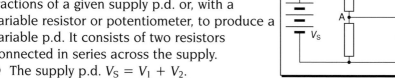

- The supply p.d. $V_S = V_1 + V_2$.
- The output p.d. $V_1 = V_S \times \dfrac{R_1}{R_1 + R_2}$
- The current in both resistors is the same when there is no current in the connection A (The only condition you need to know about).
- The ratio R_1/R_2 equals the ratio V_1/V_2. With equal resistors the supply p.d. is halved. When R_1 is very small, V_1 is very small. When R_2 is very small, V_1 approximately equals V_S.
- The potentiometer behaves as a fixed resistor between its ends but has a sliding contact, which enables you to alter the ratio of R_1 to R_2.

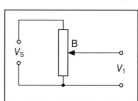

Check yourself

Electric circuits

1 State the values of the currents I_1, I_2 and I_3 in the first circuit on the right. Explain your reasoning. (3)

2 State the values of the p.d.s V_1 and V_2 in the second circuit on the right. Explain your reasoning. (2)

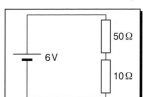

3 A battery of emf 6.0 V supplies a load of resistance 4.0 Ω. The p.d. across the battery is 5.6 V. Calculate the internal resistance of the battery. (3)

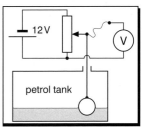

4 A lamp has a resistance of 2.5 Ω at its normal operating temperature. Two cells are available to power it. Both cells have an emf of 1.60 V. Cell A has an internal resistance of 0.5 Ω and cell B 2.5 Ω. For each cell, calculate the current in the lamp and the power dissipated in the lamp. Comment on the usefulness of each cell as a power source for the lamp. (7)

5 A potential divider is made from a 50 Ω resistor in series with a 10 Ω resistor connected as shown to a 6.0 V supply. Calculate:

(a) The current in the resistors. (2)

(b) The p.d. across each resistor. (2)

6 The diagram shows a potentiometer connected in a circuit to measure the level of petrol in a car petrol tank. The sliding contact is moved by a float in the tank. The voltmeter measures the p.d. between the sliding contact and the positive terminal of the battery. Explain how the scale of the voltmeter can be calibrated to register the level of petrol in the tank. (4)

The answers are on pages 109–110.

Sensors

Sensors are devices that provide an electrical output dependent on some other property such as temperature. Where the electrical output is a change in resistance the sensor is usually part of a potential divider.

A sound sensor (the crystal microphone)

A **piezo-electric** crystal is one which, when mechanically deformed, generates a small p.d. When sound waves strike a piezo-electric crystal the pressure changes squash and stretch the crystal so that it generates a p.d. that follows the pressure changes in the sound wave. The output waveform is an electrical analogue of the sound waveform.

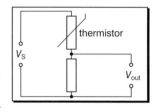

A temperature sensor (a thermometer)

The **NTC thermistor** (see page 42) is a resistor whose resistance decreases as temperature increases. The thermistor is usually incorporated in a potential divider circuit so that the output is a p.d. related to temperature. In the circuit shown the thermistor is the top resistor and the output

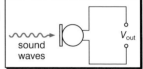

p.d. is taken across the lower resistor. When the thermistor is hot its resistance is small so that the output p.d. is large; when it is cold the output p.d. is small. (Think about how the p.d. of the supply V_S is divided between the two resistors. The smaller the top resistor relative to the lower resistor, the larger V_{out} is as a fraction of V_S.)

The relationship between temperature and output p.d. is not linear because the resistance varies logarithmically with temperature. The thermistor thermometer needs careful calibration.

A light sensor

In a **light-dependent resistor** (**LDR**) (see page 44) the resistance decreases with increase in light intensity. Like the thermistor it is usually used in a potential divider circuit as the top resistor. The output p.d. is large when the light intensity is high, giving a low resistance for the LDR.

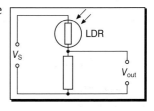

The variation of resistance with light intensity is logarithmic rather than linear, so the output voltage is not linearly related to the light intensity.

Superconductivity
(Required for AQA(B) only)

The resistance of most metals falls to a very low value at very low temperatures. In a few cases, the resistance suddenly drops to zero at a transition temperature specific to that metal. Below this transition temperature the metal is said to be **superconducting**.

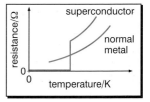

In the superconducting state very large currents can be used because no energy is dissipated in heating the metal when $R = 0$. The limiting current is set by the magnetic field generated by the current. Pure metals have transition temperatures below 20 K but some alloys and compounds have been found with transition temperatures as high as 90 K (liquid air temperature).

Superconducting alloys are used in large electromagnets (for example in body scanners) and in some energy storage devices because no heat is generated in the windings, although the whole assembly needs to be cooled to a very low temperature.

The magnetic effects of a current
(Required for OCR(A) only)

Whenever a charge moves there is a **magnetic field** associated with it. The magnetic field is represented by **field lines** or **flux lines**.

● The magnetic field strength is given by the **magnetic flux density** (*B*).
● All flux lines are closed loops.
● On diagrams, a larger field is represented by the flux lines being closer together.
● On diagrams, ⊗ (the tail of the arrow) represents a flux line or current going into the page, and ⊙ (the point of the arrow) a flux line or current coming out of the page.

Magnetic field patterns

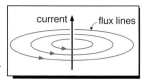

A long straight wire
● The flux lines are circles going round the wire.
● The direction of the flux is given by the **right-hand grip rule**.

A solenoid or electromagnet
● The flux lines are closed loops which thread the solenoid.
● The flux density inside the solenoid can be increased by increasing the current, the number of turns per metre or by mounting the solenoid on an iron core.

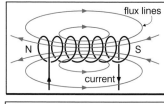

A flat circular coil
● The coil behaves as a very short solenoid.

Look at the end of a coil or solenoid. When the current is clockwise you are looking at a South pole, when the current is anti-clockwise you are looking at a North pole (see diagram at top of next page).

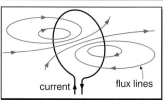

Electromagnetic forces

When a wire carrying a current is placed in a
magnetic field, at right angles to the field, it
experiences a force. The direction of this force

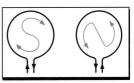

is perpendicular to both the current and the magnetic field. The force
F is given by:

$$F = BIl$$

- B is the magnetic flux density
- I is the current in the wire
- l is the length of wire in the magnetic field.

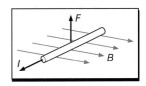

The relative orientations of B, F and Il are given
by the **left-hand rule**:

- thuMb for the *force* or *Motion* produced
- First or index finger for the *magnetic Field*
- seCond finger for the *Current*.

When the magnetic field arises from a current in
a wire this rule is equivalent to **parallel
currents attract, anti-parallel currents repel**.

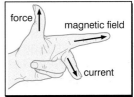

Electric motors make use of this force. In the
simple d.c. motor, the current in wires wound on
the rotor interacts with the magnetic field of the
stationary field magnet to produce a torque
which turns the rotor.

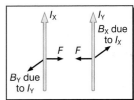

The unit of magnetic flux density (tesla)

This electromagnetic force leads to the definition of a unit for
magnetic flux density. The equation above can be rearranged to give
magnetic flux density $B = \frac{F}{Il}$.

- Magnetic flux density is measured in **tesla (T)**. The units are
  $N\,A^{-1}\,m^{-1}$.

Alternating currents
(Required for AQA(A) only)

The 'mains' electricity supply is an alternating current (a.c.) supply. The generators produce currents and p.d.s which vary with time. The graph of p.d. against time is a sine wave with the equation $V = V_0 \sin(2\pi f t)$ where f is the frequency of the supply.

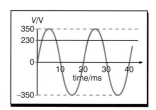

From the graph it is clear that, over one cycle, the average p.d. is zero because the positive and negative halves of the wave are equal. What is needed is an average value of the alternating p.d. which is equivalent to p.d. in a d.c. circuit.

For an alternating p.d. we define:
- V_0 is the **peak** p.d., the maximum positive or negative value.
- V_{rms} is the **root mean square (rms) p.d.** $= \frac{V_0}{\sqrt{2}}$.

Similarly, for an alternating current $I_{rms} = \frac{I_0}{\sqrt{2}}$.

The rms current and p.d. are defined in this way so that an a.c. supply delivers the same power as a d.c. supply when rms values of current and p.d. are used.
- For the British 'mains' supply $V_{rms} = 230\,V$. The peak p.d. $V_0 = 325\,V$ and the peak-to-peak p.d. $= 650\,V$ ($+325\,V$ to $-325\,V$).
- A load rated at 5 A means $I_{rms} = 5\,A$, giving a peak current of 7 A.
- The frequency of the 'mains' supply is 50 Hz. One complete cycle of the supply takes 20 ms.

Measuring a.c. currents and p.d.s
- A.c. ammeters and voltmeters are calibrated to read rms values.
- The cathode ray oscilloscope (CRO) is a convenient way of measuring a.c. voltages. From its display you can find the peak p.d. V_0 and the frequency of the supply. From V_0 you can calculate V_{rms}.
- To measure a current using a CRO, measure the p.d. across a resistance R carrying the current and calculate the current using $I_0 = \frac{V_0}{R}$.

Sensors, magnetism and alternating currents

1 A thermistor thermometer consists of a potential divider made up of the thermistor and a 500 Ω resistor connected across a 6.0 V supply. A voltmeter is connected across the 500 Ω resistor. From the graph to the right find:

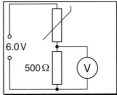

(a) The resistance of the thermistor at 20°C and at 100°C. (2)

(b) The reading on the voltmeter at these two temperatures. (3)

2 An LDR is used in the potential divider circuit shown to switch on a light at dusk. Under these conditions the LDR has a resistance of 680 Ω. Calculate the resistance of the other resistor for the voltmeter to read 4.5 V at dusk. (3)

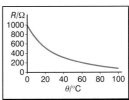

3 A wire carrying a current of 15 A lies perpendicular to a magnetic flux density of 0.25 T.

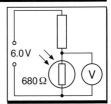

(a) Calculate the force on a 30 cm section of the wire. (2)

(b) Draw a diagram to show the direction of the force. (1)

4 The horizontal component of the Earth's magnetic field has a value of 18 μT from south to north. Calculate the force on 1.0 m of wire carrying a current of 6.0 A in an east to west direction. (3)

5 The diagram shows a simple d.c. electric motor. State in which direction it will rotate and explain how it works. (4)

6 A washing machine is connected to a 230 V rms, 50 Hz supply. It draws 3.0 kW from the supply. Calculate the rms current and peak current drawn from the supply. (3)

The answers are on pages 110–111.

(**No** work on waves is required for Edexcel, only **refraction** for AQA(A))

A large number of phenomena exhibit wave properties; for example, ripples moving on the surface of water, sound, light and other electromagnetic waves.

- Progressive waves **transfer energy** from source to receiver.
- The **energy** transferred is proportional to the **square of the amplitude**.
- Sound, water and other mechanical waves do not transfer the medium through which they move.
- Electromagnetic waves do not require a medium, they involve oscillations of electric and magnetic fields.

When you look at a water wave on the surface of a ripple tank you see water moving up and down as the wave moves across the surface. To describe this wave we can draw two graphs, **A** of displacement of the water against position at any one time (a 'photograph' of the wave), or **B** of displacement against time at any one position (the motion of an individual particle).

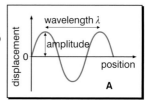

- The **wavelength** of the wave (λ) is the distance between two peaks, or any two identical positions on the wave (graph **A**).
- The period of the wave (τ) is the time for the oscillation to repeat (graph **B**).

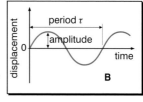

- The **frequency** of the wave (f) is the number of waves or oscillations per second.
- The unit of **frequency** is the **Hertz (Hz)**. 1 Hz = 1 wave per second.
- The frequency and period are related by **frequency = 1/period**, $f = 1/\tau$.
- The **speed of propagation** of the wave = **frequency × wavelength**,
$$V = f\lambda$$

Wave fronts and rays

On a photograph of water waves you could draw in equally spaced lines along the wave crests – the **wave fronts**. Alternatively, you could draw lines, called **rays**, perpendicular to the wave fronts showing the direction of energy transfer. Both conventions are used.

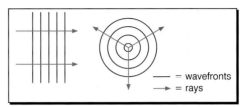

—— = wavefronts
→ = rays

Transverse waves

In a ripple tank the water moves up and down as the wave moves across the surface; in **electromagnetic waves** the electric and magnetic fields oscillate at right angles to the

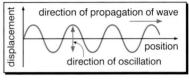

direction of the wave. These waves are called **transverse waves**, because the displacement is **perpendicular to the direction of propagation** of the wave.

Longitudinal waves

Sound waves are pressure waves. The particles of the medium move **parallel to the direction of propagation** to create the changes in pressure. Such waves are called **longitudinal waves**.

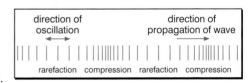

Polarisation

All transverse waves can be **polarised**. In unpolarised light the electric and magnetic fields oscillate in all directions perpendicular to the direction

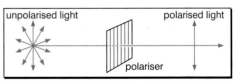

of propagation. When unpolarised light is polarised its electric and magnetic fields each oscillate in one direction only. Longitudinal waves can not be polarised.

Reflection and refraction of waves

When drawing wave or ray diagrams:
- wave fronts are drawn perpendicular to rays
- for wave fronts, the angles of incidence i and of reflection or refraction r are measured relative to the reflecting or refracting surface
- for rays, the angles of incidence and of reflection or refraction are measured relative to the **normal**
- the **normal** is a line perpendicular to the surface at the point of reflection or refraction.

When a wave meets a change in medium, in most cases some of its energy is reflected and some is refracted.

Reflection

The **laws of reflection**:
- The **angle of incidence i equals the angle of reflection r**.
- The incident ray, the reflected ray and the normal are all in the same plane.

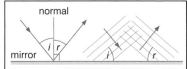

Refraction

When a wave moves from one medium to another, unless the wave meets the interface head on ($i = 0$), the wave front or ray is bent.
- The wave speeds up or slows down at the interface because **waves travel at different speeds in different media**.
- When the wave slows down the ray is bent towards the normal ($i > r$).
- When the wave speeds up the ray is bent away from the normal.
- When the wave changes speed the **frequency remains constant** and the wavelength changes.

The **laws of refraction**:
- The ratio $\frac{\sin i}{\sin r}$ **is a constant** called the **refractive index** (this is called Snell's law when applied to light).
- The incident ray, the refracted ray and the normal are in the same plane.

The refractive index n is the ratio of the speeds of the waves in the media.

- For light going from free space (a vacuum) into any medium:

$$\text{refractive index} = \frac{\textbf{(speed of light in free space)}}{\textbf{(speed of light in medium)}} = \frac{c}{c_m}$$

- For light going from free space (a vacuum) to air there is almost no change in speed, so $n_{air} = 1.00$. Free space n_0 is exactly 1.
- For light going from air to water, $_{air}n_{water} = 1.33$ because light in water travels at 0.75 of its speed in air.
- For various types of glass n varies but is approximately 1.5.
- Going from medium 1 of refractive index n_1 to medium 2 of refractive index n_2, the relative refractive index $_1n_2 = \frac{n_2}{n_1}$. (e.g. going from water, $n_1 = 1.33$, to glass, $n_2 = 1.50$, has a relative refractive index of $\frac{1.50}{1.33} = 1.13$. Going from glass to water has a relative refractive index of $\frac{1.33}{1.50} = 0.89$).

Total internal reflection

Total internal reflection (TIR) can occur when light travelling in a medium of high refractive index, such as glass, meets a boundary with a medium of lower refractive index, such as air.

Total internal reflection is so called because:
- all the light energy is reflected (**total** reflection)
- the reflected light stays **internal** to the higher refractive index medium
- the reflected light obeys the laws of **reflection**, not refraction.

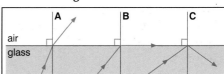

When the angle of incidence
- is small, the light is partly refracted and escapes (figure A)
- equals the **critical angle**, the light is refracted parallel to the surface (figure B)
- is greater than the critical angle, the light is totally internally reflected. (Figure C).

The critical angle C is given by the equation $\sin C = \dfrac{1}{_{air}n_g}$

- For air to water, $C = 48.7°$. For air to glass $C \approx 42°$.

Fibre optics

One application of TIR is in fibre optics. A very thin glass or plastic fibre is clad with a material of lower refractive index. Light entering the end of the fibre meets the side of the fibre

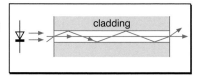

at an angle of incidence of nearly 90° and is totally internally reflected. It is thus propagated down the fibre until it escapes at the far end.

● Light travelling down the centre of the fibre arrives at the receiver in a shorter time than light reflected many times off the sides (light which has taken a longer path). This is called **smearing** and is minimised by using very fine **monomode** fibres.

● Light energy is absorbed in the fibre, resulting in **attenuation** of the signal. This is minimised by using very pure glass for the fibre.

Lenses

(Required for CCEA, Edexcel(SH) and OCR(B) only)

When curved refracting surfaces are used light may be converged (focused) or diverged (spread out). For A level you need only consider converging lenses.

● The **focal length** f of a lens is the distance from the lens to the image of a distant object.

● The **power** of a lens $= \dfrac{1}{\text{focal length}}$. It is measured in **dioptres** (D) or m^{-1}.

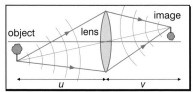

● **Converging** (**convex**) lenses like the magnifying glass, a camera lens and the lens in your eye have **positive powers**.

● For a thin converging lens the focal length f, the distance u of the object from the lens and the distance v of the image from the lens are related by: $\frac{1}{u} + \frac{1}{v} = \frac{1}{f}$ on the real-is-positive convention

(CCEA, Edexcel (SH)), or $\frac{1}{v} = \frac{1}{u} + \frac{1}{f}$ on the new cartesian convention (OCR(B)).

● The linear magnification produced by a thin lens is $\frac{v}{u}$.

Check yourself

Waves

1 The note A on the musical scale has a frequency of 440 Hz. Calculate its wavelength:
   (a) In air. (2)
   (b) In water. (1)

   (speed of sound in air = 330 m s⁻¹, in water = 1400 m s⁻¹)

   (speed of sound in air = $330\,m\,s^{-1}$, in water = $1400\,m\,s^{-1}$)

2 (a) Explain the difference between *transverse* and *longitudinal* waves. (2)
   (b) Classify the following waves as transverse or longitudinal: radio, infrared, sound, sea waves, X-rays. (1)

3 (a) State what is meant by a *polarised* wave. (1)
   (b) Explain why the signal picked up by a UHF television set changes when the aerial is rotated so that the rods are vertical rather than horizontal. (2)

4 The diagram shows a wave moving through medium 1 and about to enter medium 2, in which the wave speed is half as great. Draw in the continuation of the wave as it moves through medium 2. (2)

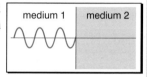

5 The diagram shows a ray of light striking a glass block of refractive index 1.50 at an angle of incidence *i* of 55°. It is partially reflected and partially refracted. Calculate the angles, r_1 and r_2, of reflection and refraction. (4)

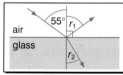

6 A glass fibre has a refractive index of 1.55 and is clad with a material with a refractive index of 1.50. Calculate the critical angle for light to be totally internally reflected within the fibre. (3)

7 A camera is fitted with a lens of focal length 50 mm.
   (a) Calculate the power of the lens. (1)
   (b) The camera is used to photograph an object 2.50 m from the lens. Calculate the distance from the lens to the film for a sharp image. (2)

The answers are on page 111.

When two wave systems overlap the two waveforms **superpose** to give a resultant waveform whose value, at any point, is the sum of the values of the component waves at that point. In the diagram, waves A and B superpose to give C.

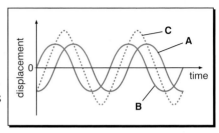

Phase difference

The **phase difference** between **two waves** is expressed as a fraction of a complete wave. For two waves with the same amplitude and frequency:

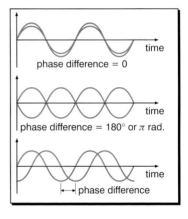

- When the waves are in step (or in phase) the phase difference is zero.
- When there is a difference of one complete wave the waves are in step again; one complete wave corresponds to a phase difference of **360° or 2π rad**.
- When a peak coincides with a trough the phase difference is **180° or π rad**.

Interference of waves

Two wave systems with the same frequency and a constant phase relationship are said to be **coherent**. When two coherent wave systems superpose they produce an **interference** pattern.
- When two coherent waves are **in phase** they interfere **constructively**.
- Constructive interference of sound waves results in a loud sound, of light waves a bright region.
- When two coherent waves are **out of phase by 180°** (π **rad**) they interfere destructively.
- Destructive interference of sound waves results in quietness, destructive interference of light waves results in a dark region.

Diffraction

When waves pass through a slit of width comparable to their wavelength they spread out (**diffract**).

- The narrower the slit, the more the waves spread out or diffract.
- The diffracted waves from different parts of the slit superpose to give a characteristic pattern of light and dark (maxima and minima) on a screen remote from the slit.

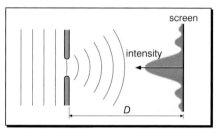

Young's double slit interference experiment

Coherent light (usually a laser nowadays) illuminates two narrow slits in a screen. Light coming through the slits diffracts. Where the diffracted beams overlap, interference is observed. A screen placed in this region shows light and dark patches of light called 'fringes'. The 'fringes' move further apart when:

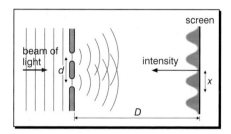

- the screen is moved further from the slits, or
- the separation of the slits is reduced, or
- light of a longer wavelength is used.

The equation relating the wavelength λ, the separation d of the slits and the slit-to-screen distance, D, to the fringe separation x is

$$x = \lambda D/d$$

Interference occurs because light reaching the screen from each of the slits has travelled a different distance. When this difference in path is zero or a whole number of wavelengths the waves are in step and superpose to give a maximum amplitude (**constructive interference**).

When the path difference is $\frac{1}{2}$, $\frac{3}{2}$, etc. of the wavelength the phase difference is 180° and the waves superpose to give a minimum (**destructive interference**).

Young's double slit experiment can be used to measure the wavelength of the waves used. For this the formula is rearranged to give $\lambda = \frac{xd}{D}$.
- To get good values d must be of the same order of magnitude as λ and D must be much greater than d.
- To find λ for microwaves, use two slits 0.10 m apart in a large sheet of metal. Detect the 'fringes' with a microwave detector + CRO.
- To find λ for sound, use two small loudspeakers 1 m apart, fed from the same sound generator. Detect the 'fringes' by a microphone + CRO.

The diffraction grating

(Required for AQA(B), CCEA and OCR(B) only)

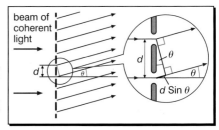

The **diffraction grating** consists of a large number of **equally spaced slits**. When the slits are illuminated with a parallel beam of coherent light interference of light from all the slits occurs in the region beyond the slits. Since the slits are equally spaced, the path difference between any pair of slits is the same. The condition for constructive interference for light from all the slits (a bright 'fringe') is

$$d \sin \theta = n \lambda$$

where d is the separation of any pair of slits, λ is the wavelength of the light, n is an integer, and θ is the angle between the normal and the direction of the 'fringe'.

The diffraction grating provides an accurate method of measuring the wavelength of light.

(OCR(B) uses an alternative argument leading to the condition for constructive interference maxima in interference patterns. Consult the Student's Book or CD for this treatment.)

Standing waves

Two coherent progressive waves travelling in opposite directions interfere to form a **standing wave**. In the diagrams you can see the two waves, one moving right and one moving left. When the two waves are superposed, sometimes they give a larger wave which does not move, and sometimes they cancel out.

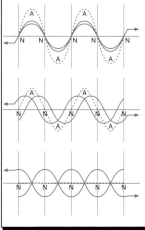

- At certain places (N, N) the waves always add up to zero. These places are called **nodes** and here the wave amplitude is always a **minimum**.
- The wavelength of the standing wave is twice the distance between two nodes.
- At places (A, A) the amplitude adds up to a **maximum**. These places are called **antinodes**.
- The nodes and antinodes are fixed in space. The **standing wave does not move** even though the progressive waves setting it up do move.
- The energy associated with a standing wave is 'lumped' around the antinodes and does not move along.
- The energy stored in a wave is proportional to the square of the wave amplitude.

The easiest way to set up a standing wave is to reflect a progressive wave back along its original path. Here are some examples of standing waves.

Standing waves on a stretched string

A vibrator sends waves along the string which are reflected from the fixed end. At certain frequencies a standing wave is set up. This is the principle of stringed musical instruments such as the guitar and the violin.

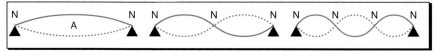

For a string fixed at both ends there must be nodes at both ends. The simplest standing wave will have one antinode and is called the **fundamental** of the string.

● The wavelength of the standing wave is twice the length l of the string.
● Other standing waves, called overtones, have wavelengths of l, $\frac{2l}{3}$, $\frac{l}{2}$ ...
● The frequencies and wavelengths of the standing waves are related by $f\lambda = v$, where v is the speed of waves along the string.
● For a given material, v is greater the thinner the string and the greater the tension.

Standing waves in air in a tube

A loudspeaker sends sound waves down a tube and these are reflected from the end. This is the principle of musical instruments like the flute and the trumpet.

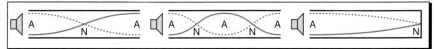

● A **closed** end of a tube is a displacement **node**.
● An **open** end of a tube is a displacement **antinode**.
● A tube with a **closed** end has a fundamental standing wave with a wavelength of $4l$ and overtones of wavelength $\frac{4l}{3}$, $\frac{4l}{2}$ ...
● A tube with an **open** end has a fundamental standing wave with a wavelength of 2l and overtones of wavelength l, $\frac{2l}{3}$, $\frac{l}{2}$ ...
● The presence of nodes and antinodes may be observed by placing fine powder in the tube. At nodes the powder stays still, at antinodes the particles of the powder jump about in the tube.

Measuring the speed of sound in air

Most methods depend on producing short pulses of sound and measuring the time taken for the sound pulse to reach a reflector and for the echo to return to a detector near the source. The formula required is $v = \frac{s}{t}$, where s is the distance from source to reflector and back to the detector, and t is the time between producing the pulse and the echo returning.

Check yourself

Superposition of waves

1 Explain what is meant by the superposition of waves. Draw the resultant wave from superposing the two waves shown. (4)

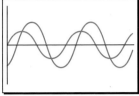

2 Explain what is meant by *constructive* and *destructive* interference. Describe how you would set up an experiment to demonstrate interference. (5)

3 Light of wavelength 550 nm illuminates two slits 0.75 mm apart. An interference pattern is observed on a screen 1.80 m from the slits. Calculate the separation of two bright fringes on the screen. (3)

4 Two loudspeakers connected to the same oscillator and emitting sound in phase with each other are mounted 1.2 m apart. At a point which is 3.40 m from one speaker and 3.65 m from the other speaker, no sound is detected. Calculate the largest wavelength possible for the sound from the speakers. (3)

5 A diffraction grating with 3.0×10^5 slits per metre is illuminated by a beam of light of wavelength 620 nm. Find the smallest angle to the beam at which a bright maximum will be observed. (3)

6 A microwave transmitter is set up opposite a metal mirror. Standing waves of wavelength 30 mm are observed between the transmitter and the mirror. State the distances between:

(a) Adjacent nodes. (1)

(b) A node and its adjacent antinode. (1)

7 An ultrasonic 'tape measure' sends pulses of ultrasound across a room and records the time taken for the echo pulse to return. In a particular room, the echo pulse returns after 17 ms. Calculate the distance from the instrument to the wall of the room. (Speed of ultrasound in air = 330 m s^{-1}.) (2)

The answers are on page 111.

Electromagnetic radiation is neither a wave nor a particle, it is electromagnetic radiation. However, to explain its behaviour we give it both **wave-like** and **particle-like** behaviour. Remember which set to use for each problem.

Wave-like properties include frequency, wavelength, speed of propagation and rate of energy transfer. The wave model enables us to explain the propagation, reflection, refraction and polarisation of electromagnetic wave energy. Superposition of waves enables us to work out the distribution of wave energy when electromagnetic waves are diffracted and interfere.

Particle-like properties include mass, momentum, speed and kinetic energy. The particle model enables us to explain the interaction of electromagnetic radiation with matter.

The family of electromagnetic waves

All electromagnetic waves travel at the same speed in free space. This speed c is $3.0 \times 10^8 \, \text{m s}^{-1}$ and is one of the fundamental constants of nature. (In other media they do not all travel at the same speed; see refraction, page 58.)

Name	Wavelength/m in free space	Frequency/Hz
gamma rays	10^{-16} to 10^{-10}	10^{18} to 10^{24}
X-rays	10^{-13} to 10^{-8}	10^{16} to 10^{21}
ultraviolet	10^{-8} to 4×10^{-7}	8×10^{14} to 10^{16}
visible	4×10^{-7} to 7×10^{-7}	4×10^{14} to 8×10^{14}
infrared	7×10^{-7} to 10^{-3}	10^{11} to 4×10^{14}
microwaves	$10^{-3} \times 10^{-1}$	3×10^9 to 10^{11}
radio waves: VHF and UHF	0.1 to 3	10^8 to 3×10^9
radio waves: SW, MW, LW	3 to 10^6	300 to 10^8

The members of the family are classified according to their wavelength or frequency. You should know their names and approximate wavelengths.

Intensity and the inverse square law

When you **double the distance** from a source of electromagnetic radiation the intensity drops to **one-quarter of its value**. This is the conservation of energy as, when a source twice as far away, the energy radiated by the source is spread out over four times the area. This is called the **inverse square law**.

Photons

When light, or any other form of electromagnetic radiation, interacts with matter the energy is transferred in 'quanta' or 'lumps' called **photons**.
● Photons all travel at the speed of light c in the medium.
● Photons carry energy $E = hf$ where f is the frequency of the radiation and h is the Planck constant.
● The higher the frequency (the shorter the wavelength) the more energy each photon carries.

The Planck constant h

The Planck constant is the constant of proportionality relating energy to the frequency of the electromagnetic radiation. Its value is very small. It is the linking factor between the wave and the particle models of matter.

$$h = 6.6 \times 10^{-34} \text{ J s}$$

Photon energies

Photon energies are very small fractions of a joule. They are often measured in electron volt (eV) rather than joule. (1 eV is the energy gained by 1 electronic charge when falling through a potential difference of 1 volt.)

$$1 \text{ electron volt} = 1 \text{ eV} = 1.6 \times 10^{-19} \text{ J}$$

Energies of X-rays and gamma rays are often measured in MeV, mega-electron volt (1 MeV $= 1.6 \times 10^{-13}$ J).

Here are some typical photon energies:
- For visible light $E = 2$ to 3 eV or 3×10^{-19} to 5×10^{-19} J
- For gamma rays $E = 0.1$ to 10 MeV or 10^{-14} to 10^{-12} J
- For infrared from bodies near room temperature $E = 0.02$ eV or 10^{-21}J.

Quantum phenomena explained using the photon model include:
- the photoelectric effect
- the light-emitting diode (LED)
- line spectra (radiation emitted from excited atoms).

The photoelectric effect

When light of certain wavelengths illuminates a clean metal surface electrons are emitted from the surface. This is called the **photoelectric effect**.

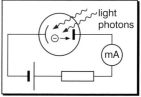

For a given metal in the photocell illuminated with monochromatic light (light of a single wavelength or colour):
- when the wavelength is greater than a threshold value there is *no* photoelectric current
- the threshold value of the wavelength is a characteristic property of the metal
- when the wavelength is shorter than the threshold value, the greater the intensity of the light and the greater the photoelectric current.

Explaining the photoelectric effect

Electrons do not usually escape from the surface of a metal. To escape work must be done on the electron to pull it out of the surface. The energy to do this work comes from the electromagnetic radiation.

The **wave model** would suggest that wave energy would be spread uniformly over the metal surface. After switching on there would be a wait until any one electron gained enough energy to escape, and then a large number would escape. In practice, electrons are emitted promptly however weak the illumination. Hence the predictions of the wave model do not agree with experimental observations.

The **quantum (photon) model** suggests that the radiation consists of a stream of photons. Each photon interacts with a single electron, giving all its energy to that electron. The total energy hf of the photon is used to remove the single electron from the metal and give that electron kinetic energy.

The Einstein photoelectric equation

Conservation of energy for a photon releasing an electron from a metal gives

$$hf = \phi + (E_k)_{max}$$

- ϕ is the 'work function', the energy required to release the electron from the metal.
- $(E_k)_{max}$ is the maximum kinetic energy of the electrons that are released from the metal.

The photon model provides an explanation which agrees with observation.

- When $hf < \phi$ no electrons are released because the photon cannot give the electron enough energy to escape. There is a threshold minimum frequency for the effect to be observed.
- Greater illumination means more photons per second, and hence more photoelectrons per second (i.e. a greater photoelectric current).
- For a given photon energy ($> \phi$) the maximum kinetic energy of the electrons released is constant, irrespective to the intensity of illumination.

The light-emitting diode (LED)

The LED behaves like an ordinary diode in that it conducts in the forward direction when a p.d. greater than its turn-on p.d. is applied, and does not conduct in the reverse direction (see page 43).

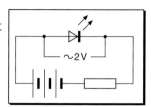

Semiconductors do not contain many free electrons so the battery has to do work to release them. Some of the freed electrons can de-excite (lose the energy given them) by emitting light. These are light-emitting diodes (LEDs).

- The energy per electron from the battery must be greater than the photon energy of the light emitted.
- Different materials will emit light of different colours because of their different energy level spacing.
- Blue LEDs require a larger forward p.d. than red LEDs because blue photons are more energetic than red photons.

Line spectra of atoms

(Required for AQA(B), CCEA, OCR(B) and WJEC only)

The electrons surrounding the nucleus of an atom are found to have well defined energies dependent on the atom under study. The allowed electron energies are represented by lines on an energy level diagram.

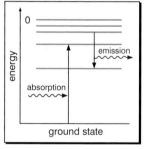

- The energy values quoted are negative and represent the energy required to remove an electron from that level to outside the atom (to **ionise** the atom).
- When an electron is **excited** it absorbs energy and moves from a lower level to a higher level. That energy could come from absorbing a photon, or from heating the atoms, or from an electric discharge.
- When an electron moves from a higher level to a lower level its excitation energy is transformed into a photon. The atom emits light.

The energy gap between the two levels can be calculated from the photon frequency using $E = hf$.

Absorption spectra are observed when white light (light of all possible wavelengths) illuminates electrons which absorb photons of just those energies that correspond to exciting electrons between the energy levels of the atoms to which they are bound.

Emission spectra are observed when light is emitted by electrons falling back from their excited state to lower states.

- **Line spectra** occur with atoms where there are discrete energy levels. The only colours emitted or absorbed are those corresponding to transitions between these levels.
- The energy levels of any atom are not equally spaced and their energy values are a 'finger print' of the atom.
- Analysis of spectra can help with identifying elements present in objects as diverse as stars and paint.

To observe line spectra, light is split up into its component frequencies in a spectrometer. The 'lines' seen are images of the narrow entrance slit of the spectrometer at the particular frequencies emitted by the source.

Check yourself

Electromagnetic waves and photons

1   State the names of the types of electromagnetic radiation labelled A and B. (2)

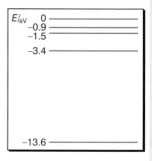

2   Calculate the energies of photons with wavelengths of:

   (a) 550 nm (green light). (1)

   (b) 3×10^{-11} m (X-ray). (1)

3   A sodium street lamp radiates yellow light of wavelength 590 nm at a power of 25 W. Calculate the number of photons produced per second. (3)

4   Explain why the simple wave model is inadequate in explaining the observations of the photoelectric effect. The longest wavelength of electromagnetic radiation that will release electrons from caesium metal is 650 nm. Calculate the work function of caesium metal. (5)

5   The work function of sodium metal is 2.5 eV. Calculate the maximum kinetic energy of photoelectrons released from sodium when illuminated by ultraviolet light of wavelength 150 nm. (4)

6   In a particular circuit a red LED is measured to have a p.d. of 2.1 V across it. Show that this is consistent with it emitting red light of wavelength 650 nm. (3)

7   The diagram shows the energy levels of a hydrogen atom. Draw on it arrows showing:

   (a) The ionisation of an electron in the ground (lowest) state. (1)

   (b) The absorption of a photon of energy 1.9 eV. (1)

   (c) The emission of a photon of energy 10.2 eV. (1)

The answers are on page 112–113.

Electron diffraction and the de Broglie relation

(Not required for Edexcel)

Electrons were originally thought of as simple particles. However, when a beam of electrons passes through a very thin sheet of crystals of graphite the electrons are diffracted and a pattern of fuzzy concentric rings is observed. The pattern is similar to that observed when a beam of light is diffracted by a small circular hole in a screen (page 63).

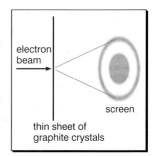

de Broglie suggested that this wave-like behaviour (diffraction) would be observed for all 'particles' and that the relation between the wave-like and particle-like behaviour was given by:

$$p = \frac{h}{\lambda}$$

where p is the momentum mv of the particle (see page 30), h is the Planck constant, and λ is the wavelength associated with the wave-like behaviour.

The de Broglie relation predicts correctly the observations of the electron diffraction experiment.

● Increasing the accelerating p.d. increases the velocity of the electrons.
● Increasing the velocity of the electrons increases their momentum.
● Using $p = \frac{h}{\lambda}$ gives a decreasing wavelength as the momentum increases.
● With a given diffracting hole, a smaller λ will give rings of smaller diameter.

Worked example

Calculate the wavelength of electrons accelerated through 2000 V.
Kinetic energy of the electrons $= Vq = (2000\ V)(1.6 \times 10^{-19}\ C)$
$$= 3.2 \times 10^{-16}\ J.$$
The velocity of these electrons $= \sqrt{(2\frac{E}{m})} = \sqrt{\frac{(6.4 \times 10^{-16}\ J)}{(9.1 \times 10^{-31}\ kg)}}$, which equals $2.65 \times 10^7\ m\,s^{-1}$.

The momentum p of these electrons is
$$mv = (9.1 \times 10^{-31}\,kg)(2.65 \times 10^7\,m\,s^{-1}),$$
which equals $2.4 \times 10^{-23}\,N\,s$.
Using de Broglie's equation $p = \frac{h}{\lambda}$ and substituting gives
$$\lambda = \frac{h}{p} = \frac{(6.6 \times 10^{-34}\,J\,s)}{(2.4 \times 10^{-23}\,N\,s)} = 2.7 \times 10^{-11}\,m.$$

To observe diffraction effects we need a hole of diameter comparable to λ for the electron. The spacing of carbon atoms in graphite is $3.7 \times 10^{-10}\,m$ so these electrons have a wavelength about 0.1 of the lattice spacing. Hence significant diffraction of the electrons will be observed.

de Broglie's equation can be applied to all 'particles'.
- Photons always travel at the speed of light so the momentum p of the photon is defined to be $\frac{E}{c}$. Using $c = f\lambda$ and $E = hf$, this gives $p = \frac{h}{\lambda}$ for the photon.
- The wavelengths of electrons with energies up to a few keV are comparable to the dimensions of atoms. Thus electron diffraction experiments give information on the structure of crystals and of atoms (the electron microscope makes use of electron diffraction).
- The wavelengths of more massive particles with much higher energies (e.g. alpha particles of 1–10 MeV) are very much smaller (about $10^{-15}\,m$). Scattering experiments with these particles give information on the structure of the nucleus.
- A student running along the road has a de Broglie wavelength. However, it is so small that there is no possibility of observing wave-like behaviour of the student!

(The ideas of electron 'waves' give rise to the wave equations representing standing electron waves in atoms. In these the square of the amplitude of the wave gives the probability of finding the electron at that point.)

The Döppler effect

(Required for AQA(B) and Edexcel(SH) only)

You will almost certainly be aware of the Döppler effect from observations on sound waves.

- The **Döppler effect** is observed only when there is **relative motion between the source and the observer.**
- The greater the relative speed, the greater the shift in frequency.
- When a source of sound **approaches** the observer the apparent frequency is **raised.**
- When the source of sound **recedes** from the observer the apparent frequency is **reduced.**

For **sound waves** a source of sound coming towards you (e.g. a police car siren) appears to have a higher pitch than when it is stationary. When the sound is going away from you, the pitch sounds lower.

For **light waves** a source of light (e.g. a star) approaching the observer (e.g. on Earth) will have the frequencies of its spectral lines shifted towards the blue end of the spectrum. A source of light receding from the observer will have the frequencies of its spectral lines shifted towards the red end of the spectrum.

To calculate the **Döppler shift** we use the relation

$$\frac{\Delta f}{f} = \frac{v}{c}$$

where Δf is the change in frequency produced by the relative motion, f is the original frequency of the source, v is the relative velocity of source and observer, and c is the speed of light (or sound).

Note: this relation applies only when $v \ll c$.

A terrestrial use of the Döppler effect with electromagnetic waves is the radar speed 'gun'. The device measures the shift in frequency of waves reflected from the moving vehicle compared with the waves transmitted by the stationary source. Here $\frac{\Delta f}{f} = 2\frac{v}{c}$ because the reflection of the source in the car is moving towards the detector at twice the velocity v of the car.

The expanding universe

(Required for AQA(B) only)

When measurements are made of the Döppler shift in light from distant galaxies, it is found that:
- the shift is always towards lower frequencies (a red shift) implying that the galaxies are receding from Earth
- the more distant the galaxy, the greater the shift in frequency.

These two results are evidence for the **expanding universe**. The age of the universe can be estimated from extrapolating backwards in time to the **big bang** – before matter had condensed out and started to expand.

Hubble's law

Hubble used the red shift to calculate the velocity v of recession of galaxies and related that to the distance d away of the galaxies. He found the simple law

$$v = Hd$$

where H is called the Hubble constant. Its value is approximately $3 \times 10^{-18} \, s^{-1}$.

Measurement of the Hubble constant enables an estimate to be made of the age of the Universe. The time taken for a galaxy, receding from us at velocity v, to get to a distance d away is $\frac{1}{H}$ assuming that all matter was clumped together at the 'big bang'. $\frac{1}{H} \sim 3 \times 10^{17}$ s, which makes the age of the Universe $\sim 10^{10}$ years.

Check yourself

de Broglie, Döppler and Hubble

1 Make a list of phenomena in which electrons show:

   (a) Particle-like behaviour. (2)

   (b) Wave-like behaviour. (2)

2 An electron has a momentum of $4.5 \times 10^{-24}\,\text{N s}$. Use de Broglie's equation to find its wavelength. (2)

3 Explain why electron diffraction is observed with crystals but not with slits of separation 0.5 mm in a metal screen. (2)

4 Estimate the wavelength of a rifle bullet of mass 15 g travelling at $350\,\text{m s}^{-1}$. Explain why diffraction effects are not observed with rifle bullets. (4)

5 A spectral line of frequency $5.0 \times 10^{14}\,\text{Hz}$ in light from a distant galaxy is found to be shifted by $3.0 \times 10^{12}\,\text{Hz}$. Calculate the velocity of recession of the galaxy. (2)

The answers are on pages 113–114.

The structure of the atom

(Required for AQA(A), AQA(B), Edexcel and Edexcel(SH), WJEC only)

The **Rutherford** or **nuclear** model of the atom has a very small dense nucleus surrounded by a cloud of electrons.
- The **atom** has a diameter of the order of 10^{-10} m, which is the diameter of the electron cloud.
- The **nucleus** at the centre of the atom has a diameter of the order of 10^{-15} to 10^{-14} m.
- The nucleus has a positive charge and is made up of two types of **nucleons** – **protons** and **neutrons**.
- Evidence for the nuclear atom comes from alpha particle scattering experiments (see page 84).

Particle	Symbol	Relative mass	Relative charge
electron	e	1/1837	−1
proton	p	1	+1
neutron	n	1	0

- The charge on the electron (*e*) has a magnitude of 1.6×10^{-19} C.
- The mass of the proton and of the neutron is 1.7×10^{-27} kg. This is also called one **atomic mass unit (u)**.

Isotopes
- Any one **element** is defined by the number of protons in the nucleus.
- That number of protons is called the **atomic number (Z)**.
- Any one element may have several **isotopes** with different numbers of **neutrons** in the nucleus.
- The total number of protons and neutrons is called the **mass number (A)**.
- The number of neutrons in the nucleus is (A–Z).

- The nucleus of a particular isotope is called a **nuclide**.

An isotope of element X is written $_Z^A$X when it has Z protons and (A–Z) neutrons. For example:
- Hydrogen (H) has three isotopes $_1^1$H (p), $_1^2$H (p + n), $_1^3$H (p + 2n).
- The uranium isotope $_{92}^{238}$U has 92 protons and 146 neutrons.

Radioactivity

(Required for AQA(B), Edexcel and Edexcel(SH) only)

Whenever a radiation counter is set up it detects some radiation even when no known sources are near. This is called **background radiation**. It is made up of:
- **cosmic radiation** from the sun and other stars
- **terrestrial background radiation** from radioactive materials in rocks, brickwork, etc. in the environment
- radioactive residues in the atmosphere as a result of burning coal, nuclear weapons testing in the 1950s and accidents such as that at Chernobyl.

In the environment a few isotopes of a small number of elements are found to emit radiation spontaneously. These isotopes are said to be naturally **radioactive** and the process is called **radioactive decay**.
- The original nucleus before decay is called the **parent** nucleus.
- The nucleus produced as a result of the decay is called the **daughter** nucleus.

The random nature of radioactive decay

When measurements are made on, for instance, background radiation the count rate is found to vary even though its average value remains constant. This is because **radioactive decay is a random process**. We cannot say when any particular nucleus will decay, just give the probability that it will decay in the next time interval.

The classification of radiation from radioactive nuclei is given in the table.

Radiation	Nature	Mass/u	Charge/e	Ionising ability
alpha (α) particles	helium nuclei	4	+2	very strong
beta (β) particles	electrons	1/1837	−1	strong
gamma (γ) rays	electromagnetic radiation	0	0	very weak

Individual radioactive nuclides decay by the emission of (α only) and (α followed by γ), (β only) and (β followed by γ), but **not** (α followed by, or following, β) or (γ only).

The absorption of radiation

All types of radiation (α, β and γ) lose energy as they ionise atoms when passing through matter. The production of ionisation is the basis of the Geiger–Muller radiation detector and of the health hazards associated with radiation.

Alpha particles
● Create a lot of ionisation in matter due to their large mass and charge.
● Typically have high energies in the range 1–5 MeV.
● From any one nuclide all usually have the same energy, and hence the same range in matter.
● Have a range of roughly 10 cm in air or 0.25 mm in water or tissue.

Beta particles
● Create quite a lot of ionisation in matter as they are electrons.
● Typically have maximum energies in the range 100 keV to 1 MeV.
● From any one nuclide have a range of energies up to a maximum, and hence no well defined range in matter.
● Have a range of about 30 cm in air or 3 mm in aluminium.

Gamma rays
● Are not strongly ionising as they are uncharged electromagnetic radiation.
● Typically have energies in the range 50 keV to 3 MeV.

- From any one nuclide have certain well defined energies like line spectra.
- Are attenuated exponentially in passing through matter. (There is a thickness that will halve their intensity but not totally absorb them.)
- In air show the inverse square law. (When you move the counter twice as far away from the source the count rate drops to one quarter.)

Below are absorption curves for radiation in a homogeneous medium:

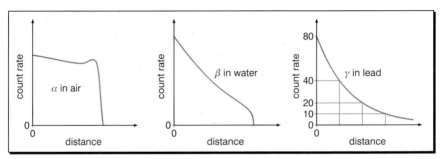

Balancing nuclear equations

When a radioactive nucleus decays both charge and number of nucleons (protons + neutrons) are conserved.

- The total Z on the parent must equal the total Z on the products.
- The mass number A of the parent must equal the total of the mass numbers of the products. For example:

(α decay) $^{238}_{92}U$ $^{234}_{90}Th + ^{4}_{2}α + γ$ (energy)

Z: 92 for U = 90 for Th + 2 for α. A: 238 for U = 234 for Th + 4 for α. γ has no charge or mass.

(β decay) $^{90}_{38}Sr$ $^{90}_{39}Y + ^{0}_{-1}β$

Z: 38 for Sr = 39 for Y + (−1) for β. A: 90 for Sr = 90 for Y + 0 for β.

The half-life of a radioactive isotope

The activity or rate of decay of a radioactive isotope is proportional to the number of undecayed nuclei remaining.

● The unit of activity is the **Becquerel** (Bq). 1 Bq = 1 decay per second.

● For any given isotope there is a constant time, the **half-life**, for the **activity to halve** in value.

● A graph of activity against time has a characteristic shape (an exponential shape) showing the constant half-life.

In any measurement, the time taken for the count rate to halve is constant. From an initial count rate of $1000\,s^{-1}$ there will be count rates of $500\,s^{-1}$ ($\frac{1}{2}$) after 1 half life, $250\,s^{-1}$ ($\frac{1}{4}$) after two half lives, $125\,s^{-1}$ ($\frac{1}{8}$) after three half lives, ... $1\,s^{-1}$ ($\frac{1}{1024}$) after ten half lives.

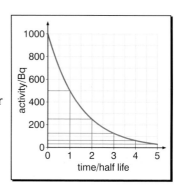

Evidence for the nuclear atom

Rutherford, Geiger and Marsden fired a beam of α-particles at a very thin gold foil. The whole apparatus was mounted in a high vacuum system. Measurements of the angles through which the α-particles were deflected were consistent with the model of a very small dense nucleus at the centre of the atom.

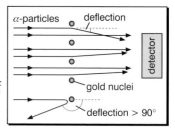

The measurements showed that:

● Most of the α-particles went through the foil **undeflected**, suggesting the atom was mostly empty space.

- A **few** α-particles were deflected through small angles.
- A **very few** α-particles were deflected through **more than 90°**, suggesting that a massive nucleus scattered those few α-particles that hit it.
- An α-particle that recoiled (deflected through about 180°) must have approached to within 10^{-14} m of the centre of the nucleus, suggesting that the nucleus was very small (less than 10^{-4} of the diameter of the atom).

Elastic and inelastic scattering

Bombarding nuclei with various particles is the principal way of determining the structure of nuclei. Different particles and different energies are used to study different properties.

Elastic scattering occurs when the bombarding particle suffers no change in kinetic energy in the scattering. The α-particle scattering experiment above is an example of elastic scattering at low energy (1–5 MeV).

Inelastic scattering occurs when the bombarding particle loses energy to the bombarded nucleus. High-energy (1 GeV) inelastic scattering of electrons gives information on the structure of the nucleus and of neutrons and protons.

Radioactivity

1 State how many protons and neutrons there are in the nucleus of $^{32}_{15}P$. (2)

2 The radioisotope $^{238}_{92}U$ decays through a chain of α-decays and β-decays to the stable isotope $^{206}_{82}Pb$. Calculate how many α-particles and how many β-particles are emitted in the complete decay process.(4)

3 The radioisotope $^{60}_{27}Co$ decays by the emission of a β-particle and two γ-rays. Write a balanced nuclear equation for the decay.(4)

4 Oxygen (atomic number = 8) has three stable isotopes with mass numbers 16, 17 and 18. List the numbers of electrons, protons and neutrons in neutral atoms of these isotopes of oxygen. (4)

5 Explain what is meant by the term half-life as applied to the decay of a radioactive isotope. A sample of sodium–24 (^{24}Na) has an activity of 4000 Bq. The half-life of sodium–24 is 15 hours. Calculate the activity of the sample after 75 hours. (4)

6 The graph shows how the count rate of a radioactive source varies with time over a period of 500 s. Use the graph to find three values of the half-life of the radioactive source. (3)

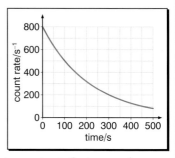

7 Explain how Rutherford's α-particle scattering experiment supports the model of the nuclear atom. Gold was used as the target because very thin gold foil (leaf) was available. Why is it important that a nucleus with large values of Z and A is used? (5)

The answers are on page 114.

(Required for AQA(A), AQA(B) and Edexcel topic 3C)

Bombarding matter with particles such as electrons and protons of high energies gives information on the structure of the nucleus and its components. In particular, collision experiments using very high energies show that the proton and the neutron have structure and are not fundamental particles.

There is a large number of sub-nuclear particles, which can be classified into various groups.

Antiparticles

Every particle has an associated **antiparticle**. For example, the antiparticle of the **electron** β^- is the **positron** β^+.
- The positron has the same mass as the electron.
- The positron has a charge of $+e$ ($e = 1.6 \times 10^{-19}$ C), whereas the electron has a charge of $-e$.
- The positron annihilates with an electron when they collide to give two γ-rays: $_{-1}^{0}\beta^- + _{+1}^{0}\beta^+$ 2γ.

When any particle and antiparticle meet they **annihilate** each other. The annihilation interaction must conserve mass/energy* and momentum. That is why two γ-rays are produced rather than one in the electron–positron annihilation. Each γ-ray takes half the energy and they go off in opposite directions, showing that momentum is conserved.

*Mass/energy is written because mass and energy can be interconverted using Einstein's equation $\Delta E = c^2 \Delta m$. A useful conversion is that 1 u $= 930$ MeV.

Pair production

When a high-energy γ-ray passes near a nucleus it may vanish and create an electron–positron pair. To do so the energy of the γ-ray must be greater than the energy equivalent of the rest masses of the electron and positron.

The classification of particles

Leptons

The electron is a member of a group of particles called **leptons**.
- Leptons are fundamental particles and have no structure themselves.
- There are six members (and their antiparticles) in the group, as shown in the table.

particle	symbol	lepton number	rest energy/MeV
electron	β^-	+1	0.51
positron	$\beta+$	−1	0.51
electron-neutrino	υ_e	+1	0
electron-antineutrino	$\bar{\upsilon}_e$	−1	0
negative muon	μ^-	+1	105
positive muon	μ^+	−1	105
muon-neutrino	υ_μ	+1	0
muon-antineutrino	$\bar{\upsilon}_\mu$	−1	0
negative tau	τ^-	+1	1780
positive tau	τ^+	−1	1780
tau-neutrino	υ_τ	+1	0
tau-antineutrino	$\bar{\upsilon}_\tau$	−1	0

Hadrons

The proton and the neutron are not fundamental particles. They belong to the family of **hadrons**, which are made up of different numbers of **quarks**. The hadrons are split into two groups called **baryons** and **mesons**.

Baryons include the proton, antiproton, neutron and antineutron. The only stable baryon is the proton. The neutron (outside the nucleus) decays into a proton, electron and electron-antineutrino with a half-life of 11 minutes.

The proton and neutron both have a **baryon number** of +1, the antiproton and antineutron have a baryon number of −1 and mesons have a baryon number of 0.

Mesons include pions (π^-, π^+, π^0) and kaons (K^-, K^+, K^0). Mesons decay into leptons and γ-rays.

Quarks

There are six quarks and six antiquarks. They may be fundamental constituents of matter. At this level you need to know about only three of the quarks – **up**, **down** and **strange**. Their properties are given in the table (e is the charge on the electron).

quark	symbol	charge/e	baryon number	strangeness
up	u	$+\frac{2}{3}$	$+\frac{1}{3}$	0
anti-up	$\overline{u}$	$-\frac{2}{3}$	$-\frac{1}{3}$	0
down	d	$-\frac{1}{3}$	$+\frac{1}{3}$	0
anti-down	$\overline{d}$	$+\frac{1}{3}$	$-\frac{1}{3}$	0
strange	s	$-\frac{1}{3}$	$+\frac{1}{3}$	-1
anti-strange	$\overline{s}$	$+\frac{1}{3}$	$-\frac{1}{3}$	-1

Baryons are made up of three quarks (or three antiquarks):
- A proton is (u u d) and an antiproton ($\overline{u}\ \overline{u}\ \overline{d}$).
- A neutron is (u d d) and an antineutron ($\overline{u}\ \overline{d}\ \overline{d}$).

Mesons are made from one quark and one antiquark:
- For example, (π^+ is (u $\overline{d}$), K^0 is (d $\overline{s}$) etc.

Exchange forces

The forces that hold particles together, or force them apart, are thought of in terms of the exchange of particles.

Imagine two skaters, A and B, moving alongside each other on parallel paths (a). When A throws a heavy ball towards B he recoils and moves away from B (remember the conservation of momentum) (b). When B catches the ball it changes his direction away from A (conservation

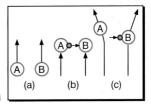

of momentum again) (c). The ball models the exchange particle and the exchange of the ball is a model for a repulsion force.

Feynman diagrams

These exchange forces ideas were formalised by Richard Feynman into a set of simple diagrams to represent the interactions. In the Feynman diagram:
- time starts at the bottom and moves from bottom to top
- straight lines represent the interacting particles
- wavy lines represent the exchange particles.

For **electrical forces** the exchange particle is the photon (γ) so the diagram for the repulsion between two electrons looks like the top diagram here.

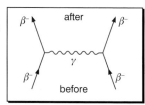

For β^- decay the exchange particle is the W^- particle, so the diagram for β^- decay looks like the bottom diagram.

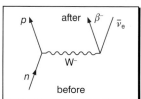

The weak nuclear force

The weak nuclear force is responsible for β decay. It involves both leptons and quarks. The decay of the neutron into a proton (β^- decay) is represented by the equation

$$n \qquad p + \beta^- + \bar{\upsilon}_e.$$

Note that an electron-antineutrino is required to balance the equation so that the total lepton number remains the same (zero in this case).

The number of baryons is the same (1) before and after, but a d quark has changed into a u quark.

The corresponding equation for positron decay is

$$p \qquad n + \beta^+ + \upsilon_e.$$

The strong nuclear force

The strong nuclear force between quarks holds protons and neutrons together in the nucleus. Its exchange particle is the **gluon**.

Conservation laws

With all these particles as building blocks there is a large number of interactions that could happen, although only a few have been observed. Those observed fit a set of rules called **conservation laws**.

- **Charge, mass/energy, lepton number** and **baryon number** are always conserved. That is, the total after the interaction is over is the same as the total before the interaction took place.
- The conservation of mass/energy implies that the interacting particles must have enough energy to create the rest mass of any products created.
- **Strangeness** is conserved in strong nuclear interactions but can change by 0 or ± 1 in weak nuclear interactions.

The conservation laws enable us to identify those interactions which happen when there is sufficient energy available. For example:

1 Electron capture

$$p + \beta^- \qquad n + \upsilon_e.$$

Applying the conservation laws gives

charge	$(+e) + (-e)$	$0 + 0$ ✔
lepton number	$0 + (+1)$	$0 + (+1)$ ✔
baryon number	$(+1) + 0$	$(+1) + 0$ ✔

All laws satisfied, so electron capture is observed.

2 Will we observe this decay?

$$n \qquad \pi^0 + \gamma.$$

Applying the conservation laws gives

charge	0	$0 + 0$ ✔
lepton number	0	$0 + 0$ ✔
baryon number	$(+1)$	$0 + 0$ ✘

Baryon number is not conserved, so this decay will not be observed.

Check yourself

Particle physics

In these questions you may use the following data: $1u = 930\,\text{MeV}$, the rest mass of an electron $= 0.51\,\text{MeV}$.

1 Use the table of the properties of quarks on page 89 to show that a proton can be made up from two up and one down quark. (2)

2 State the quark composition of the neutron. Describe β^- decay in terms of quarks and draw a Feynman diagram to represent β^- decay. (4)

3 **(a)** Calculate the minimum energy of a single γ-ray that can create an electron–positron pair. (2)

(b) Calculate the energies of the γ-rays produced when a low-energy positron annihilates with a stationary electron. (2)

4 A particle accelerator accelerates protons to an energy of $5.0\,\text{GeV}$. They collide head on with antiprotons, also of energy $5.0\,\text{GeV}$ moving in the opposite direction. Each collision produces an additional proton–antiproton pair. Calculate what fraction of the total energy of the original pair is required to create the additional pair. (4)

5 State, giving your reasons, the quark composition of π^0 and π^+ mesons. (4)

6 Use the conservation laws to predict whether this weak nuclear interaction decay of a muon can happen:

$$\mu^- \qquad \beta^- + \overline{\upsilon}_e + \upsilon_\mu \ (4)$$

7 Describe the interaction represented by the Feynman diagram to the right. (4)

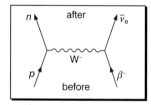

The answers are on page 114.

(Required for AQA(A), Edexcel and Edexcel(SH) only)

Temperature

Temperature, the hotness of a body, may be defined from the idea that thermal energy flows from hot to cold.
- The **celsius** temperature ($\theta°$C) is defined in terms of $0°$C and $100°$C, the freezing and boiling points of water at 1 atm pressure.
- The **kelvin temperature** (T K) is the absolute temperature scale starting at 0 K as the absolute zero of temperature.
- The two scales are related by $\frac{T}{K} = \theta°$C $+ 273$
- The temperature intervals are the same on both scales: $\Delta\theta$ in celsius equals ΔT in kelvin.

Thermal energy measurements

When thermal energy (heat energy) is supplied to a body its temperature rises. The thermal energy necessary to produce a rise in temperature of 1 K depends on the mass (m) of the material and its **specific heat capacity** (c), which is a constant for the material.
- The specific heat capacity is the energy, in joule, required to raise 1 kg of the material by K (or 1 °C)
- The unit for specific heat capacity is $J\,kg^{-1}\,K^{-1}$.
- The thermal energy E supplied or removed is given by Energy = mass × specific heat capacity × change in temperature. $E = mc\Delta\theta$.

When a material changes state – melts or freezes, vaporises or condenses – energy transfers take place at constant temperature. The energy transfer per kilogram is called the **specific latent heat** and is measured in $J\,kg^{-1}$.
- The thermal energy E supplied or removed in a change of state is given by Energy = mass × specific latent heat. $E = mL$
- The latent heat associated with freezing or melting is called the **latent heat of fusion**.
- The latent heat associated with vaporisation or condensation is called the **latent heat of vaporisation**.

The behaviour of gases

The equation of state for an ideal gas is

$$pV = nRT$$

This equation is a combination of the following:
- Boyle's law (pV is constant at constant temperature)
- $\frac{V}{T}$ is constant at constant pressure
- $\frac{p}{T}$ is constant at constant volume.

The last two of these relations suggest that if $V \propto T$ and $p \propto T$ there is a zero of temperature which has a physical significance, the absolute zero of temperature. This is taken to be the zero for the kelvin scale.

In the equation $pV = nRT$
- p is the pressure in Pa, V the volume in m^3 and T the temperature – which **must** be measured in **kelvin**.
- n is the amount of gas measured in mole.
- R is the universal gas constant, $R = 8.3\,\text{J}\,\text{mol}^{-1}\,\text{K}^{-1}$.
- It does not matter which gas is used as long as the amount of gas is measured in mol.

Many gases such as oxygen, hydrogen and helium behave as 'ideal' gases over the range of p, V and T encountered in many applications around room temperature.

The kinetic theory of an ideal gas

The kinetic theory models a gas as a large number of particles rattling around at random in a container. Direct evidence for the model comes from observing the **Brownian motion** of pollen grains in water or smoke particles in air.

Assumptions about the particles forming the ideal gas:
- They are small hard spheres that occupy negligible volume (the gas can be compressed into a negligible volume).
- They exert negligible forces on each other, except when they collide (the gas particles do not stick together).

- They collide elastically so that kinetic energy is conserved in collisions (the gas does not cool spontaneously).

The pressure in a gas

When the particles of the ideal gas collide with the walls of the container their momentum is changed. Force is rate of change of momentum, so the change in momentum of particles colliding with the walls provides a force on the walls and hence a pressure inside the container.

- The pressure p is given by $p = \frac{1}{3}\rho \overline{c^2}$, where ρ is the density of the gas and $\overline{c^2}$ is the mean square speed of the particles (the mean of the squares of the speeds of all the particles).
- When the temperature is raised the average speed of the particles increases and the pressure rises because the particles hit the walls more often and there is a greater change in momentum (greater force) from each collision.

Internal energy and temperature

The **internal energy (U)** of an ideal gas is the kinetic energy of the random motion of its particles.

- The internal energy of an ideal gas is proportional to its kelvin temperature.
- For one mole of gas the internal energy is $\frac{3}{2}RT$
- The kinetic energy E_k of one particle is $\frac{1}{2}m\overline{c^2} = \frac{3}{2}kT$ where $k = \frac{R}{N_A}$. This is known as the Boltzmann constant (for one particle). N_A is the Avogadro constant, the number of particles in one mole.

For real gases and other states of matter the internal energy is the sum of the kinetic and potential energies of the particles making up the system (the energy internal to the boundary of the system).

Thermal equilibrium

The kinetic theory implies that there is continual interchange of energy in collisions between particles. Two bodies are said to be in **thermal equilibrium** when the rate of transfer of thermal energy from one to the other equals the rate of transfer from the other to the one. This is called a **dynamic equilibrium**.

Thermal physics

1 An electric kettle heats 0.50 kg of water from 20 °C to 100 °C. Calculate the energy required. (Specific heat capacity of water $= 4200 \, \text{J} \, \text{kg}^{-1} \, \text{K}^{-1}$.) (2)

2 The electric kettle heater in question 1 is rated at 2000 W. Calculate the time taken for the water in the kettle to boil. (2)

3 The electric kettle's cut-out fails and the 2000 W heater continues to vaporise the water into steam. Calculate the mass of water evaporated in one minute. (Specific latent heat of vaporisation of water $= 2.2 \, \text{MJ} \, \text{kg}^{-1}$.) (3)

4 A student places a beaker of cold water at 8 °C in a freezer. Explain why cooling the water to 0 °C takes much less time than freezing the water into ice at 0 °C. (Specific heat capacity of water $= 4200 \, \text{J} \, \text{Kg}^{-1}$, specific latent heat of fusion of ice $= 330 \, \text{kJ} \, \text{Kg}^{-1}$.) (4)

5 For a fixed mass of an ideal gas, sketch graphs of:

(a) Pressure against volume at constant temperature. (1)

(b) Pressure against temperature (in K) at constant volume. (1)

(c) Volume against temperature (in K) at constant pressure. (1)

6 A weather balloon is filled with $5.0 \, \text{m}^3$ of helium at a pressure of 100 kPa and a temperature of 290 K. It rises to where the pressure is 30 kPa and the temperature is 240 K. Calculate the new volume of the helium in the balloon. (3)

7 Explain, in terms of the kinetic theory of gases, the observation that the volume of a gas increases when the pressure is reduced at constant temperature (Boyle's law). (3)

8 Calculate the kinetic energy of a helium atom at 300 K and hence the root mean square speed of helium atoms at 300 K. (Boltzmann constant $= 1.4 \times 10^{-23} \, \text{J} \, \text{K}^{-1}$, mass of helium atom $= 6.7 \times 10^{-27} \, \text{kg}$.) (4)

The answers are on pages 115–116.

(Required for AQA(B) and OCR(B) only)

Data collected from sensors is usually in the form of an electrical potential difference. This must be captured, stored and processed for later display.

Analogue and digital data

An **analogue** signal is one in which the potential difference varies continuously and can have any value in its range. The output of many sensors is analogue. For example, the output of a crystal microphone is the electrical analogue of a sound waveform. The outputs of all the sensor circuits on pages 48–49 are analogue outputs.

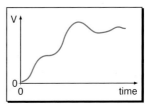

A **digital** signal can only have one of two values (usually labelled '0' and '1'). Different information comes from coding a series of '0' and '1' pulses.

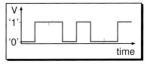

Digital images

A **digital image** is made up of an array of dots or elements called **pixels**. Each pixel is represented by a number giving its location, colour and brightness. The picture shows the letter P on an array of 8 × 6 pixels, each of which is either black or white. Because the pattern is stored as a number rather than as a complete letter it can be manipulated and displayed in many forms.

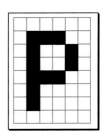

The datalogger

The datalogger captures information from sensors and stores it digitally for subsequent processing and display.

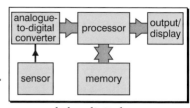

- With analogue sensors an **analogue-to-digital** converter has to be placed between the input from the sensor and the digital processor.
- The quality of the information captured depends on the number of sensors. For example, the **resolution** or quality of a digital photograph depends on the **number of pixels** on the CCD (light detector) chip in the camera.
- When the information changes with time the quality depends on the **sampling rate**: the more samples taken during the time of the experiment, the more information is available for analysis.

Digital signals

One digital or binary pulse, a '0' or a '1', is called one **bit** of information. A group of 8 bits is called one **byte**.

- One byte is an 8-bit number which can have 256 values. Thus, the output of a pixel coded as an 8-bit number could represent the light falling on the pixel as 256 different colours (or 256 shades of grey in a black and white system).
- The resolution of a digital signal depends on the number of bits used to code each quantity. An 8-bit code gives 256 values for the quantity, a 16-bit code gives 65 000 values.
- Digital signals are easy to store and process in computers.
- Digital signals can be **multiplexed** to enable several sets of digital information to be transmitted down a line at the same time. (Each set of information is sampled in rotation and its value transmitted. When this is done quickly enough, all the information appears to be transmitted simultaneously.)
- Digital signals can be recovered when attenuated or noisy because only the sequence of pulses is required, not the pulse shape or amplitude.

Analogue signals
- Analogue signals cannot be stored or processed directly in modern computers. They must first be digitised.
- Noise and interference cannot be removed from analogue signals when they are amplified.

Transmitting information

Information may be transmitted as:
- electrical signals along wires and cables
- visible light and infrared along optical fibres
- radio and microwave propagation in the atmosphere.

In all three systems the information is degraded by noise and interference, and attenuated (energy is dissipated) as the distance between transmitter and receiver is increased.

Transmission system	Attenuation process	Noise and interference
Metal conductors	The wires have resistance and dissipate the energy of the pulses	Noise arises from thermal effects; interference is picked up from adjoining circuitry
Optical fibres	Absorption of light in the glass; minimised by using very pure glass	Little noise and no interference picked up from adjoining fibres
Radio and microwaves	Absorption and scattering in the atmosphere	Noise and interference from other transmitters picked up by the detector

Sound signals

Audible sound ranges in frequency from 50 Hz to 15 kHz. For any particular sound we can plot a **frequency spectrum** (a plot of the amplitudes of the various frequencies which superpose to make the sound). To transmit 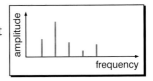 music we need to sample all the frequencies in the audible range. Sampling is usually done at 40 kHz with 16-bit resolution. Therefore, one second of music uses 640 kbit of information; a transmission rate of 640 000 bit s^{-1}.

Voice telephones use only 8-bit resolution and a sampling rate of 8 kHz, so telephone messages require a transmission rate of only $64 \, \text{kbit s}^{-1}$.

Video signals

A picture contains much more information than a sound wave. A typical television picture has over half a million pixels, each of which has to be coded for intensity and colour 25 times per second. One video signal at $25 \, \text{Mbit s}^{-1}$ can be compared for transmission with over two hundred telephone calls.

Transmission by modulation of a carrier frequency

Information is encoded for transmission by modulating a carrier wave. This modulation may be by changing the amplitude of the carrier wave (AM) or the frequency of the carrier wave (FM).

- The carrier frequency must be much greater than the modulating frequency or bit rate.

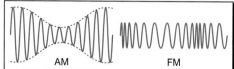

- The **bandwidth** of the system indicates the amount of information which the system can carry.
- The maximum bit rate is related to the bandwidth of the system.

Rate of transmission of information

Carrier	Frequency	Maximum bit rate
Coaxial cable	$1 \times 10^9 \, \text{Hz}$	$70 \, \text{Mbit s}^{-1}$
Microwave beam	$1 \times 10^{10} \, \text{Hz}$	$140 \, \text{Mbit s}^{-1}$
Optical fibre	$2 \times 10^{14} \, \text{Hz}$	$5000 \, \text{Mbit s}^{-1}$

Check yourself

Information and data collection

1 Write out the binary signal encoded by the train of pulses on the right. (2)

2 Explain why most information is now processed in a digital form. (4)

3 Draw a block diagram for a datalogger. Explain the purpose of each block. (5)

4 A digital camera has a CCD detector with 2.1 megapixels. The output of each pixel is coded as an 16-bit number. Calculate the number of bytes necessary to store one picture in the digital camera. (3)

5 Show that recording a music CD with 16-bit resolution and a 40 kHz sampling rate generates information for storage at 640 kbits s^{-1}. (2)

6 Use the table on page 99 to estimate how many television channels can be transmitted down an fibre optic link. Suggest why your answer is an overestimate. (3)

7 State and explain two advantages of using a fibreoptic link in preference to a pair of copper wires. (4)

The answers are on pages 116–117.

Check yourself answers

1 **(a)** 2500 m $= 2.5 \times 10^3$ m. (1) 2500 equals 2.5×1000.
  (b) 4.5 MJ $= 4.5 \times 10^6$ J (1) Prefix M means $\times 10^6$.
  (c) 2.5 mA $= 2.5 \times 10^{-3}$ A (1). Prefix m means $\times 10^{-3}$.

2 **(a)** Energy or work equals (force $\times$ distance). Force equals (mass $\times$ acceleration). Thus force (newton) $= \mathrm{kg\,m\,s^{-2}}$. (1)

  (b) Pressure equals force per unit area $=$ newton/metre^2. (1)
  From (a) the newton equals $\mathrm{kg\,m\,s^{-2}}$,
  so pressure $= \dfrac{(\mathrm{kg\,m\,s^{-2}})}{(\mathrm{m^2})} = \mathrm{kg\,m^{-1}\,s^{-2}}$. (1)

3 **(a)** $\frac{2}{3} = 0.667$ to 3 sig. figs. 1)
  (b) $\pi = 3.14$ to 3 sig. figs. (1)
  (c) The answer on the calculator is 35672. To 3 sig. figs the answer is 35700 or 3.57×10^4 in standard form. (1)

4 **(a)** $\sin \theta =$ (opposite/hypotenuse) $= 0.385$. (1)
  (b) $\cos \theta =$ (adjacent/hypotenuse) $= 0.923$. (1)
  (c) $\tan \theta =$ (opposite/adjacent) $= 0.417$. (1)

5 **(a)** Divide both sides by R to give $I = \dfrac{V}{R}$. (1)

  (b) Divide both sides by g to give $\dfrac{s}{g} = \dfrac{1}{2}t^2$;

  multiply each side by 2 to give $\dfrac{2s}{g} = t^2$. (1)
  Hence $t = \sqrt{(2\frac{s}{g})}$. (1)

  (c) Transfer c to the other side to give $y - c = mx$. Hence $x = \dfrac{(y-c)}{m}$. (1)

6 Rearrange $V = IR$ to give $R = \dfrac{V}{I}$. (1)

  Combine with $R = \dfrac{\rho l}{A}$ to get $\dfrac{V}{I} = \dfrac{\rho l}{A}$. (1)

  Rearrange this equation to get $\rho = \dfrac{VA}{Il}$. (1)

VELOCITY, ACCELERATION AND MOTION   (page 11)

1 The resultant vector is 1 when the two vectors are antiparallel (1) and 3 when they are parallel (1).

2 Draw the vector diagram. The two velocity vectors are perpendicular to each other. By calculation, ground velocity $= \sqrt{[(1.2\,\mathrm{m\,s^{-1}})^2 + (0.5\,\mathrm{m\,s^{-1}})^2]}$ $= 1.3\,\mathrm{m\,s^{-1}}$, at an angle of 67° to the flow of the river. (3)

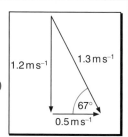

3 Data given are $s = 105\,\mathrm{km} = 1.05 \times 10^5\,\mathrm{m}$,
  $t = 1$ hour $= 3600\,\mathrm{s}$. (1)
  The formula required is $v = \dfrac{s}{t}$. Substituting in formula gives

  $v = \dfrac{(1.05 \times 10^5\,\mathrm{m})}{(3600\,\mathrm{s})} = 29\,\mathrm{m\,s^{-1}}$. (1)

4 The formula required is $a = \dfrac{(v-u)}{t}$. (1)

Check yourself answers

Data given are $u = 0$, $v = 10\,\mathrm{m\,s^{-1}}$, $t = 1\,\mathrm{s}$.

Substituting in formula gives $a = \dfrac{(10\,\mathrm{m\,s^{-1}})}{(1\,\mathrm{s})} = 10\,\mathrm{m\,s^{-2}}$. (1)

5 The formula required is $a = \dfrac{\Delta v}{\Delta t}$. (1)

Data given are $\Delta v = 30\,\mathrm{m\,s^{-1}}$, $\Delta t = 6.0\,\mathrm{s}$.

Substituting in formula gives $a = \dfrac{(30\,\mathrm{m\,s^{-1}})}{(6.0\,\mathrm{s})} = 5.0\,\mathrm{m\,s^{-2}}$. (1)

6 (a) Average speed $= \dfrac{s}{t} = \dfrac{(4000\,\mathrm{m})}{(90\,\mathrm{s})} = 44\,\mathrm{m\,s^{-1}}$. (1)

(b) Average velocity $= 0$ because velocity is a vector and the total displacement over the lap is zero. (2)

7 (a) OA: the object moved at a constant velocity away from the origin. (1) AB: the object remained at rest at a distance from the origin. (1) BC: the object returned to the origin at a slower velocity than it left along OA. (1)

(b)

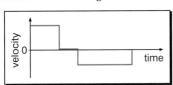

8 Constant acceleration gives constant gradient for first 8.0 s. Total journey time is 23 s. (4)

9 Total distance $=$ area of triangle $+$ area of rectangle.

Distance travelled $= \frac{1}{2}(15\,\mathrm{m\,s^{-1}})(10\,\mathrm{s}) + (15\,\mathrm{m\,s^{-1}})(30\,\mathrm{s})$ (2)

$= 75\,\mathrm{m} + 450\,\mathrm{m} = 525\,\mathrm{m}$. (1)

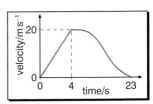

THE EQUATIONS OF MOTION   (page 15)

1 (a) The formula required is $v^2 = u^2 + 2as$. (1)

Data given are $u = 0$, $v = 2.5 \times 10^6\,\mathrm{m\,s^{-1}}$, $s = 35\,\mathrm{mm}$. Substituting in

formula gives $a = \dfrac{v^2}{2s} = \dfrac{(2.5 \times 10^6\,\mathrm{m\,s^{-1}})^2}{0.070\,\mathrm{m}} = 8.9 \times 10^{13}\,\mathrm{m\,s^{-2}}$. (1)

(b) The formula required is $t = \dfrac{s}{v}$ where $s = 0.40\,\mathrm{m}$. Substituting in formula

gives $t = \dfrac{(0.40\,\mathrm{m})}{(2.5 \times 10^6\,\mathrm{m\,s^{-1}})} = 0.16\,\mathrm{\mu s}$. (1)

2 (a) The formula required is $v^2 = 2as$ since $u = 0$. (1)

Data given are $a = 0.15\,\mathrm{m\,s^{-2}}$, $s = 75\,\mathrm{m}$. Substituting in formula gives $vs = 2(0.15\,\mathrm{m\,s^{-2}})(75\,\mathrm{m})$. $v = 4.7\,\mathrm{m\,s^{-1}}$. (2)

(b) Time $t = \frac{v}{a}$ (1) $= \frac{(4.7\,\mathrm{m\,s^{-1}})}{(0.15\,\mathrm{m\,s^{-2}})} = 31\,\mathrm{s.}$ (1)

3 The police car catches the stolen car when they have both travelled the same distance from the start. (1) The distance travelled equals the area between the graph line and the time axis. (1) Work out the area for each car and show that it is 810 m for both cars after 27 s. (1)
For the stolen car, distance travelled $= (30\,\mathrm{m\,s^{-1}})(27\,\mathrm{s}) = 810\,\mathrm{m.}$ (1)
For the police car, distance travelled $= \frac{1}{2}(36\,\mathrm{m\,s^{-1}})(9\,\mathrm{s}) + (36\,\mathrm{m\,s^{-1}})(18\,\mathrm{s}) = 162\,\mathrm{m} + 648\,\mathrm{m} = 810\,\mathrm{m.}$ (1)

4 Use the formula $v = u + at$, where $a = -g = -9.8\,\mathrm{N\,kg^{-1}}.$ (1)
On the way up $u = 300\,\mathrm{m\,s^{-1}}$ and at the top $v = 0$. Substituting gives
$t = \frac{u}{g} = \frac{(300\,\mathrm{m\,s^{-1}})}{(9.8\,\mathrm{N\,kg^{-1}})} = 30.6\,\mathrm{s.}$ (1)

On the way down t also $= 30.6\,\mathrm{s}$, therefore total time $= 61\,\mathrm{s.}$ (2)

5 (a) Vertical component velocity $= (300\,\mathrm{m\,s^{-1}})\sin 45 = 212\,\mathrm{m\,s^{-1}}.$ (1)
Horizontal component velocity $- (300\,\mathrm{m\,s^{-1}})\cos 45 = 212\,\mathrm{m\,s^{-1}}.$ (1)
(b) The formula required is $v = u + at$, where $a = -g = -9.8\,\mathrm{N\,kg^{-1}}.$ (1)
On the way up $u = 212\,\mathrm{m\,s^{-1}}$ and at the top $v = 0$. Substituting gives
$t = \frac{u}{g} = \frac{(212\,\mathrm{m\,s^{-1}})}{(9.8\,\mathrm{N\,kg^{-1}})} = 21.6\,\mathrm{s}$ up. (1)

The total time to return to ground $= 2 \times 21.6\,\mathrm{s} = 43.2\,\mathrm{s.}$ (1)
(c) The formula required is $s = vt$, where $v = 212\,\mathrm{m\,s^{-1}}$, $t = 43.2\,\mathrm{s.}$ (1)
Substituting gives $s = (212\,\mathrm{m\,s^{-1}})(43.2\,\mathrm{s}) = 9160\,\mathrm{m.}$ (2)

FORCES AND THE MOMENTS OF FORCES (page 21)

1 Acceleration of aircraft $= \frac{(v-u)}{t} = \frac{(80\,\mathrm{m\,s^{-1}})}{(15\,\mathrm{s})} = 5.3\,\mathrm{m\,s^{-2}}.$ (1)
$F = ma$ (1) $= (4.0 \times 10^5\,\mathrm{kg})(5.3\,\mathrm{m\,s^{-2}}) = 2.1 \times 10^6\,\mathrm{N.}$ (1)

2 (a) Using $v = u + at$, data given are $v = 0$, $u = 4.5\,\mathrm{m\,s^{-1}}$ and $t = 0.20\,\mathrm{s}$.
Rearranging the formula gives $a = -\frac{u}{t}.$ (1)

Substituting in formula gives $a = \frac{-(4.5\,\mathrm{m\,s^{-1}})}{(0.20\,\mathrm{s})} = -22.5\,\mathrm{m\,s^{-2}}.$ (1)

(b) Substituting in $F = ma$ gives $F = (70\,\mathrm{kg})(22.5\,\mathrm{m\,s^{-2}}) = 1600\,\mathrm{N.}$ (2)

3 (a) Weight of girl $= (55\,\mathrm{kg}) \times (9.8\,\mathrm{N\,kg^{-1}}) = 540\,\mathrm{N.}$ (1)
(b) At rest, her weight is balanced by the force upwards from the floor of the lift. Therefore the force exerted on her by the lift is 540 N. (1)
(c) To accelerate her upwards the force upwards from the floor of the lift must be greater. Formula required is $F = ma$, data given are $m = 55\,\mathrm{kg}$, $a = 1.4\,\mathrm{m\,s^{-2}}$.
Substituting in the formula gives $F = (55\,\mathrm{kg}) \times (1.4\,\mathrm{m\,s^{-2}}) = 77\,\mathrm{N.}$ (1)
Therefore total force exerted by lift $= 540\,\mathrm{N} + 77\,\mathrm{N} = 617\,\mathrm{N}$ upwards. (1)

(d)

↑ 617 N (1)

↓ 540 N (1)

4 (a) Using the formula $s = \frac{1}{2}gt^2$, data given are g and $t = 2.5\,\text{s}$. (1)
Substituting in formula gives $s = \frac{1}{2}(9.8\,\text{m s}^{-2})(2.5\,\text{s})^2 = 31\,\text{m}$. (1)
(b) The formula required is $v = gt$. (1)
Substituting in formula gives $v = (9.8\,\text{N m}^{-2})(2.5\,\text{s}) = 24.5\,\text{m s}^{-1}$. (1)

5 (a) Zero resultant force but finite resultant torque. (1)
(b) Finite resultant force and torque. (1)
(c) Zero resultant force and zero resultant torque. (1)

6 Moment of force $= (20\,\text{N}) \times (0.70\,\text{m}) = 14\,\text{N m}$. (1)

7 The centre of gravity is at the 50 cm mark (1), 0.30 m from the pivot. (1)
Balancing moments gives $(1.8\,\text{N})(0.20\,\text{m}) = (\text{weight of rule})(0.30\,\text{m})$. (1)
Hence, weight of rule $= 1.2$ N. (1)

WORK AND ENERGY (page 25)

1 (a) The formula required is work done $= \Delta E_P = mg\Delta h$.
Data given are $m = 25\,\text{kg}$, $g = 9.8\,\text{N kg}^{-1}$, $\Delta h = 1.2\,\text{m}$. (1)
Substituting gives $\Delta E_P = (25\,\text{kg})(9.8\,\text{N kg}^{-1})(1.2\,\text{m}) = 290\,\text{J}$. (1)
(b) The barrel is moved through the same vertical height so the work done is the same in each case (290 J). (1)

2 Kinetic energy $E_K = \frac{1}{2}mv^2 = \frac{1}{2}(65\text{kg})(5.5\,\text{m s}^{-1})^2 = 980\,\text{J}$. (1)
At the top of the ramp all the E_K has been transformed into E_P. (1)
Rearranging the formula gives $\Delta h = \frac{E_K}{mg}$. Substituting gives
$\Delta h = \frac{(980\,\text{J})}{(65\,\text{kg})}(9.8\,\text{N kg}^{-1}) = 1.5\,\text{m}$. (1)

3 Rearranging $E_K = \frac{1}{2}mv^2$ gives $v = \sqrt{(2\frac{E_K}{mg})}$. (1)

Substituting gives $v = \frac{\sqrt{(16 \times 10 - 13\,\text{J})}}{(6.4 \times 10^{-27}\,\text{kg})} = 1.6 \times 10^7\,\text{m s}^{-1}$. (1)

4 Loss in GPE per second $= \Delta E_P$ per second $= (m$ per second$)\,g\Delta h$. (1)
Data given are m per second $= 150\,\text{kg s}^{-1}$, $g = 9.8\,\text{N kg}^{-1}$, $\Delta h = -3.5\,\text{m}$.
Substituting gives $E_P = (150\,\text{kg s}^{-1})(9.8\,\text{N kg}^{-1})(3.5\,\text{m}) = 5100\,\text{J s}^{-1}$. (1)

Check yourself answers

Power equals energy per second, therefore the input power to the turbine is 5100 W. (1)

5 Gain in GPE $= mg\Delta h = (70\,\text{kg})(9.8\,\text{N}\,\text{kg}^{-1})(4.0\,\text{m}) = 2700\,\text{J}$. (1)

Power $= \dfrac{\text{energy}}{\text{time}}$ (1) $= \dfrac{(2700\,\text{J})}{(5.0\,\text{s})} = 550\,\text{W}$. (1)

6 Rearranging $P = Fv$ with power constant gives $F \propto \frac{1}{v}$. As the speed v rises the motive force F falls. (1) Acceleration $a = \frac{F}{m}$, so as the motive force falls the acceleration of the car is less. (1) When the motive force falls to just balance the frictional forces, the acceleration is zero and the car has reached its top speed. (1)

7 Fall in GPE $= \Delta E_P = mg\Delta h = (75\,\text{kg})(9.8\,\text{N}\,\text{kg}^{-1})(200\,\text{m}) = 150\,\text{kJ}$. (2)
Work done against friction $= Fd = (24\,\text{N})(2000\,\text{m}) = 48\,\text{kJ}$. (2)
Energy transformed into $E_K = 150\,\text{kJ} - 48\,\text{kJ} = 102\,\text{kJ}$. (1)
$E_K = \frac{1}{2}mv^2$, so $v = \sqrt{(2\frac{E_K}{mg})} = \sqrt{\frac{(204\,\text{kJ})}{(75\,\text{kg})}} = 52\,\text{m}\,\text{s}^{-1}$. At the bottom of the hill the skier is travelling at $52\,\text{m}\,\text{s}^{-1}$. (1)

FORCES AND MOTION (page 31)

1 (a) Work done $=$ force $\times$ distance (1) $= (15\,\text{N}) \times (1000\,\text{m}) = 15\,\text{kJ}$. (1)

(b) Power $= \dfrac{\text{work done}}{\text{time}}$ (1) $= \dfrac{(15\,\text{kJ})}{(60\,\text{s})} = 250\,\text{W}$. (1)

2 The force pair is the weight of the book acting downwards on your hand (1) and the force from your hand acting upwards on the book. (1)

3 The downwards force from the weight of the person in both cases. (1) The upwards drag force depends on velocity and is much larger at a given velocity with the parachute than without. (1) At the terminal velocity, the weight balances the upwards drag force. (1)

4 The weight of the object is $(2.0\,\text{kg}) \times (9.8\,\text{N}\,\text{kg}^{-1}) = 20\,\text{N}$. (1)
At $30\,\text{m}\,\text{s}^{-1}$ resultant force $= (20\,\text{N}) - (10\,\text{N}) = 10\,\text{N}$. Therefore acceleration $= \frac{F}{m} = \frac{(10\,\text{N})}{(2.0\,\text{kg})} = 5.0\,\text{m}\,\text{s}^{-2}$. (1)
At $60\,\text{m}\,\text{s}^{-1}$ resultant force $= 0$. Therefore acceleration $= 0$. (1)
As the velocity increases the acceleration decreases until it reaches the terminal velocity of $60\,\text{m}\,\text{s}^{-1}$. (1)

5 First find the deceleration using $v^2 = u^2 + 2as$. (1)
Data given are $v = 0$, $u = 30\,\text{m}\,\text{s}^{-1}$, $s = 1.6\,\text{m}$. Substituting gives

$a = \dfrac{-(30\,\text{m}\,\text{s}^{-1})^2}{(2 \times 1.6\,\text{m})} = 280\,\text{m}\,\text{s}^{-2}$. (1)

Substituting in $F = ma$ gives $F = (65\,\text{kg})(280\,\text{m}\,\text{s}^{-2}) = 18\,\text{kN}$. (2)

6 The formula relating braking distance to speed is $v^2 = 2as$. (1)
   Doubling v to $2v$ increases v^2 to $(2v)^2 = 4v^2$. (1)
   Therefore for constant a the braking distance s has increased by $\times 4$. (1)
7 Momentum $= mv = (9.1 \times 10^{-31}\,\text{kg})(2.0 \times 10^7\,\text{m s}^{-1}) = 1.8 \times 10^{-23}\,\text{N s}$. (2)
8 **(a)** Momentum of bullet $= mu = (0.01\,\text{kg})(300\,\text{m s}^{-1}) = 3.0\,\text{N s}$. (1)
   Applying conservation of momentum, $mu = (m + M)v$. (1)
   Substituting gives $3.0\,\text{N s} = (8.01\,\text{kg})v$. (1)

   (b) Velocity of box $= \dfrac{(3.0\,\text{N s})}{(8.01\,\text{kg})} = 0.37\,\text{m s}^{-1}$. (1)

9 For balls each of mass m, with A having an initial velocity v_A and B having a
   final velocity v_B, the initial velocity of B and final velocity of A both $= 0$.
   Total momentum before collision $= mv_A + (0)_B$. (1)
   Total momentum after collision $= (0)_A = mv_B$. (1)
   Conservation of momentum is satisfied when $v_A = v_B$. (1)
   Total kinetic energy before collision $= \frac{1}{2}mv_A^2$. Total kinetic energy after
   collision $= \frac{1}{2}mv_B^2$. (1)
   Kinetic energy conservation is satisfied when $v_A = v_B$. (1)

PROPERTIES OF SOLIDS   (page 35)

1 **(a)** Using the formula $F = kx$ where $F = 15\,\text{N}$ and $k = 250\,\text{N m}^{-1}$. (1)
   Substituting gives $x = \dfrac{F}{k} = \dfrac{(15\,\text{N})}{(250\,\text{N m}^{-1})} = 0.60\,\text{m}$. (1)

   (b) The formula required for the stored energy $= \frac{1}{2}Fx$. (1)
   Substituting gives stored energy $= \frac{1}{2}(15\,\text{N})(0.060\,\text{m}) = 0.45\,\text{J}$. (1)

2 **(a)** The same force provides twice the extension, so spring constant is
   halved. (1) Spring constant for two springs in series is $12.5\,\text{N m}^{-1}$. (1)
   (b) To provide the same extension, double the force is required. (1) Spring
   constant for two springs in parallel is $50\,\text{N m}^{-1}$. (1)
3 Use the formula $k = \dfrac{F}{x}$. (1) There are two springs so each supports the weight
   of 30 kg. For each spring $F = (30\,\text{kg})(9.8\,\text{N kg}^{-1}) = 290\,\text{N}$. (1)
   x for each spring is 5.0 cm $= 0.050$ m. (1)
   Substituting gives $k = \dfrac{(290\,\text{N})}{(0.050\,\text{m})} = 5800\,\text{N m}^{-1}$. (1)

4 For small loads, 20 N extended the spring by 10 mm to give a spring constant,
   the gradient of the graph, of $\dfrac{(20\,\text{N})}{(0.010\,\text{m})} = 2000\,\text{N m}^{-1}$. (2) As the
   load was increased the spring reached its elastic limit at 30 N, (1) and started
   to deform plastically at larger loads. (1) When the load was removed the
   spring had a permanent plastic deformation. (1)

Check yourself answers

5 Use the formula stress $= \frac{F}{A}$ where $F = 150\,N$ and $A = 2.4 \times 10^{-6}\,m^2$.

Substituting gives stress $= \dfrac{(150\,N)}{(2.4 \times 10^{-6}\,m^2)} = 6.3 \times 10^7\,Pa$. (2)

Using the relation strain $= \dfrac{stress}{Young\ modulus}$ (1)

Strain $= \dfrac{(6.3 \times 107\,Pa)}{(2.1 \times 1010\,Pa)} = 0.0030$ (1)

Extension = strain original length $= 0.0030 \times (5.0\,m) = 0.015\,m$. (1)

6 You must think of:
Advantage – cars deform plastically to absorb energy in crash. *Disadvantage* – when cars are dented they cannot easily be straightened. For each property (1), reasoning (1).

7 We can work out the following from the graph: the Young modulus from the gradient near the origin, the elastic limit from the bend, the ultimate tensile strength from the end and the stored energy per unit volume from the area. Young modulus $= 1.0 \times 10^{10}\,Pa$, (2) elastic limit $= 1.0 \times 10^8\,Pa$, (1) ultimate tensile stress $= 1.2 \times 10^8\,Pa$, (1) energy per unit volume $= 6.0 \times 10^6\,J$. (2)

ELECTRICITY BASICS   (page 41)

1 Using $q = It$ (1) we get $q = (3.0\,A)\,(12\,s) = 36\,C$. (1)

2 The graph starts at the origin and has the form shown. (1) The total charge transferred is calculated from the shaded area under the graph (1) This area $= \frac{1}{2}$ base $\times$ height $= 40\,C$. (1)

3 Electrons have a negative charge and will move from – to +. (1) This is in the opposite direction to the conventional current. (1)

4 The number of electrons in $1\,C$ is $\dfrac{1}{(1.6 \times 10{-}19\,C)} = 6.3 \times 10^{18}$. (1) $1\,A$ is

$1\,C\,s^{-1}$ (1) so corresponds to 6.3×10^{18} electrons passing per second. (1)

5 Using energy $W = qV$ (1) $= (2500\,C) \times (1.5\,V) = 3800\,J$. (1)

6 Using power $P = VI$ (1) $= (3.0\,V) \times (0.040\,A) = 0.12\,W$. (1)
The energy transferred is given by power time $= VIt = (3.0\,V) \times (0.040\,A) \times (3600\,s) = 430\,J$. (1)

7 Using current $I = \frac{V}{R}$ (1) gives $I = \dfrac{(6.0\,V)}{(10\,\Omega)} = 0.60\,A$. (1)

8 Rearranging $P = I^2R$ gives $I = \sqrt{(\frac{P}{R})}$. (1)

Substituting gives $I = \sqrt{\dfrac{(15\,W)}{(24\,\Omega)}} = 0.79\,A$. (1)

9 The resistance equals the gradient of the graph. Choose a point on the graph, e.g. $6.0\,V$, $503\,mA$. Gradient $= (6.0\,V)/(0.050\,A) = 120\,\Omega$. (1)

Check yourself answers

10 **(a)** In series, $R_T = R_1 + R_2$. (1) $R_T = (4\,\Omega) + (5\,\Omega) = 9\,\Omega$. (1)
(b) In parallel, $\frac{1}{R_T} = \frac{1}{R_1} + \frac{1}{R_2}$. (1) $\frac{1}{R_T} = \frac{1}{(4\,\Omega)} + \frac{1}{(5\,\Omega)} = \frac{9}{(20\,\Omega)}$. Therefore
$R_T = (\frac{20}{9})\,\Omega = 2.2\,\Omega$. (1)

ELECTRICAL RESISTANCE   (page 45)

1 Use the formula $\rho = \frac{Rl}{A}$ rearranged to give $l = \frac{RA}{\rho}$. (1)
Data given are $\rho = 1.1 \times 10^{-6}\,\Omega\,\text{m}$, $A = 6.5 \times 10^{-8}\,\text{m}^2$, $R = 2.5\,\Omega$. Substituting
gives $l = \frac{(2.5\,\Omega)(6.5 \times 10^{-8}\,\text{m}^2)}{(1.1 \times 10^{-6}\,\Omega\,\text{m})} = 0.15\,\text{m}$. (1)

2 The volume of the putty remains constant. (1) Therefore, if the length of the cylinder is multiplied by three, the area of cross-section must be divided by three. (1). $R \propto \frac{1}{A}$ so resistance is increased by nine times to be $13.5\,\Omega$. (1)

3 Current in copper wire at $20\,°\text{C} = \frac{V}{R} = \frac{(1.08\,\text{V})}{(1.40\,\Omega)} = 0.720\,\text{A}$. (1)
At $100\,°\text{C}$ the resistance is greater, therefore the current is smaller. (1)
Current at $100\,°\text{C} = \frac{(1.08\,\text{V})}{(1.98\,\Omega)} = 0.545\,\text{A}$. Current has decreased by $01.75\,\text{A}$. (1)

4 For any point on the graph, find V and I and work out resistance $= \frac{V}{I}$. (Note that the resistance is not the gradient of the graph for a non-ohmic resistor.) (1) The resistance of the filament increases as the current increases. (1)

5 Lamp A will light because the diode in series is connected in the forward direction. (1) Lamp B will not light because the diode in series is connected in the reverse direction. (1)

6 In darkness the LDR has a very high resistance. Zero on the ammeter corresponds to darkness. (1) In sunlight the LDR has a low resistance. Therefore sunlight corresponds to the maximum current reading registered by the ammeter. (1)

ELECTRICAL CIRCUITS   (page 49)

1 Use the conservation of charge. The current entering a resistor must equal the current leaving the resistor. Therefore $I_1 = I_2 = 2.0\,\text{A}$. (2)
The currents in the two resistors in parallel must add up to 2.0 A. Therefore $I_3 = 2.0\,\text{A} - 0.5\,\text{A} = 1.5\,\text{A}$ since $I_2 = 2.0\,\text{A}$. 4 (1)

2 V_1 is the p.d. across the $4\,\Omega$ resistor carrying a current of 0.5 A. Therefore $V_1 = (0.5\,\text{A})(4.0\,\Omega) = 2.0\,\text{V}$. (1)
The p.d. across the pair of resistors in series must add up to the battery p.d. Therefore $V_2 = 6.0\,\text{V} - 2.0\,\text{V} = 4.0\,\text{V}$. (1)

3 Use $I = \frac{V}{R}$ to calculate the current in the resistor.
$I = \frac{(5.6\,\text{V})}{(4.0\,\Omega)} = 1.4\,\text{A}$. (1)

Check yourself answers

To find the internal resistance the formula required is $E = V - Ir$. Data given are $E = 6.0\,V$, $V = 5.6\,V$, $I = 1.4\,A$. Substituting gives $r = \dfrac{(E - V)}{I}$
$= \dfrac{(6.0\,V - 5.6\,V)}{(1.4\,A)} = 0.29\,\Omega$. (2)

4 For cell A, current $I = \dfrac{E}{(R + r)} = \dfrac{(1.60\,V)}{(2.5\,\Omega + 0.5\,\Omega)} = 0.53\,A$. (2)

Power in lamp $= I^2R = (0.53\,A)^2(2.5\,\Omega) = 0.70\,W$. (2)

For cell B current $I = \dfrac{(1.60\,V)}{(2.5\,\Omega + 2.5\,\Omega)} = 0.32\,A$. (1)

Power in lamp $= (0.32\,A)^2\,(2.5\,\Omega) = 0.26\,W$. (1)

Although both cells have the same emf, the lamp will be much brighter when operated from cell A (the lower internal resistance) rather than from cell B. With cell A it dissipates almost three times the power. (1)

5 The resistors are in series, thus the total resistance is $(50\,\Omega + 10\,\Omega) = 60\,\Omega$. The current is the same in each resistor and equals $\dfrac{(6.0\,V)}{(60\,\Omega)} = 0.10\,A$. (2) The p.d. across the $50\,\Omega$ resistor is $IR = (0.10\ A)(50\,\Omega) = 5.0\,V$. (1) Similarly, the p.d. across the $10\,\Omega$ resistor is $1.0\,V$. (1)

6 The drawing shows that, when the tank is empty and the float is at the bottom of the tank, the sliding contact is at the positive end terminal of the resistor. (1) Thus when the tank is empty, the voltmeter reads zero. (1) As the tank is filled the slider moves up and the reading on the voltmeter increases. (1) The voltmeter can be calibrated by pouring in known volumes of petrol and marking the voltmeter scale accordingly. (1)

SENSORS, MAGNETISM AND ALTERNATING CURRENTS (page 55)

1 From the graph: $R_{20} = 550\,\Omega$, $R_{100} = 70\,\Omega$. (2)
For the potential divider use $V = (6.0\,V)(500/(500 + R))$. (1)
Substituting gives $V_{20} = \dfrac{(6.0\,V)}{(500/1050)} = 2.9\,V$. (1)
$V_{100} = \dfrac{(6.0\ V)}{(500/570)} = 5.3\,V$. (1)

2 With 4.5 V across the LDR the current I in the LDR $= \dfrac{V}{R.\,I} = \dfrac{(4.5\ V)}{(680\,\Omega)} = 6.6\,mA$. (1)
The p.d. across $R = (6.0\,V - 4.5\,V) = 1.5\,V$. (1) Resistance of $R = \dfrac{V}{I} = (1.5\,V)/(0.0066\,A) = 230\,\Omega$. (1)

3 The required formula is $F = BIl$. (1)
The data given are $B = 0.25\,T$, $I = 1.5\,A$,
$l = 0.30\,m$. Substituting gives
$F = (0.25\,T)(15\,A)\,(0.30\,m) = 1.1\,N$. (1)

4 Use $F = BIl$ with data $B = 1.8 \times 10^{-5}\,T$,
$I = 6.0\,A$, $l = 1.0\,m$.(1)
Substituting gives $F = (1.8 \times 10^{-5}\,T)(6.0\,A)(1.0\,m) = 1.1 \times 10^{-4}\,N$. (2)

5 The coil will rotate anti-clockwise when viewed from the open end of the coil. (1) Your argument will identify the direction of the magnetic field, (1) the directions of the forces on the sides of the coil (1) and the necessity of a

Check yourself answers

commutator to enable the rotation to be continuous. (1)

6 The required formula is $P = V_{rms}I_{rms}$, rearranged to give $I_{rms} = \frac{P}{V_{rms}}$. (1)

Substituting gives $I_{rms} = \frac{(3000\,W)}{(230\,V)} = 13\,A$. (1)

Peak current $I_0 = \sqrt{s}I_{rms} = 18\,A$. (1)

WAVES (page 61)

1 Use $v = f\lambda$ rearranged to give $\lambda = \frac{v}{f}$ (1)

Substituting gives (a) in air $\lambda = \frac{(330\,m\,s^{-1})}{(440\,Hz)} = 0.75$ m (1), and (b) in water
$\lambda = \frac{(1400\,m\,s^{-1})}{(440\,Hz)} = 3.2\,m$. (1)

2 You should refer to page 58 for the first part. (2)
Only sound waves are longitudinal, the rest are transverse. (1)

3 You should refer to page 58 for the meaning of polarised. (1)
The TV signal is horizontally polarised so the rods of the aerial are set horizontal. When turned vertical the rods do not pick up the TV signal. (2)

4 In medium 2 the waves should have the same amplitude (1) and half the wavelength. (1)

5 The reflected ray has angle r_1 to the normal. For reflection, angle of incidence = angle of reflection. (1) Therefore $r_1 = i = 55°$. (1)
For refraction, use Snell's law, $\sin r_2 = \frac{\sin i}{n}$, (1) substituting gives
$\sin r_2 = \frac{(\sin 55°)}{(1.50)} = 0.546$, $r_2 = 33°$. (1)

6 Relative refractive index from cladding to glass $= \frac{1.55}{1.50} = 1.033$. (1)

Formula for critical angle is $\sin C = \frac{1}{n}$. (1)

Substituting gives $\sin C = \frac{1}{1.033} = 0.968$. $C = 75°$. (1)

7 Power of lens $= \frac{1}{(focal\ length)} = \frac{1}{(0.050\ m)} = 20\,D$. (1)

Using formula relating object and image distances and focal length,

(real-is-positive) $\frac{1}{(2.50\,m)} + \frac{1}{v} = \frac{1}{(0.050\,m)}$ gives $v = 0.051$ m. (2)

(new Cartesian) $\frac{1}{v} = \frac{-1}{(2.50\,m)} + \frac{1}{(0.050\,m)}$ gives $v = 0.051$ m.

SUPERPOSITION OF WAVES (page 67)

1 Revision of superposition is on page 62. The resultant has the same wavelength, (1) same shape, (1) amplitude shown. (1)

2 The explanation of interference is given on page 62. To demonstrate interference you could describe Young's double slit experiment, a

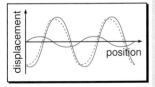

diffraction grating or standing waves. Apparatus and detail of experiment (1) diagram, (1) description of interference pattern observed (1).

3 The formula required is $x = \lambda \frac{D}{d}$. (1) Converting all lengths to m gives $\lambda = 5.5 \times 10^{-7}$ m and $d = 7.5 \times 10^{-4}$ m with $D = 1.80$ m. (1)

Substituting gives $x = \frac{(5.5 \times 10^{-7}\,\text{m})(1.80\,\text{m})}{(7.5 \times 10^{-4}\,\text{m})} = 1.33$ mm. (1)

4 The path difference between the two waves is $3.65\,\text{m} - 3.40\,\text{m} = 0.25\,\text{m}$. (1) No sound, so the point is a minimum which makes the path difference $\frac{\lambda}{2}$ for the largest value of λ . (1) Therefore $\lambda = 2 \times 0.25\,\text{m} = 0.50\,\text{m}$. (1)

5 The formula required is $n\lambda = d \sin \theta$.(1) For the first maximum $n = 1$.

Separation of slits d corresponds to $\frac{1}{(3.0 \times 105\,\text{m})} = 3.3 \times 10^{-6}$ m. (1) Substituting

gives $\sin \theta = \frac{(6.2 \times 10^{-7}\,\text{m})}{(3.3 \times 10^{-6}\,\text{m})} = 0.186$. hence $\theta = 11°$. (1)

6 The separation of 2 nodes is $\frac{1}{2}$ wavelength $= 15\,\text{mm}$. (1)

The distance from node to adjacent antinode is $\frac{1}{4}$ wavelength $= 7.5\,\text{mm}$. (1)

7 Rearranging $v = \frac{s}{t}$ gives $s = vt = (330\,\text{m s}^{-1})(0.017\,\text{s}) = 5.6\,\text{m}$. (1)

The represents the distance to the wall and back, so the wall is 2.8 m from the ultrasonic tape measure.

ELECTROMAGNETIC WAVES AND PHOTONS   (page 74)

1 A is the ultraviolet region. (1)

B is the microwave region. (1)

2 Combining $E = hf$ and $c = f\lambda$ gives $E = \frac{hc}{\lambda}$. (1)

Substituting gives (a) $E = \frac{(6.6 \times 10^{-34}\,\text{J s})(3.0 \times 10^{8}\,\text{m s}^{-1})}{(5.5 \times 10^{-7}\,\text{m})} = 3.6 \times 10^{-19}$ J and

(b) $E = \frac{(6.6 \times 10^{-34}\,\text{J s})(3.0 \times 10^{8}\,\text{m s}^{-1})}{(3.0 \times 10^{-11}\,\text{m})} = 6.6 \times 10^{-15}$ J. (2)

3 Energy of one photon $E = \frac{hc}{\lambda}$. Substituting gives $E = (6.6 \times 10^{-34}\,\text{J s})$

$(3.0 \times 10^{8}\,\text{m s}^{-1})/(5.9 \times 10^{-7}\,\text{m}) = 3.35 \times 10^{-19}$ J. (1)

Number of photons per second $= \frac{(\text{power})}{(\text{energy per photon})}$. (1)

Number $\text{s}^{-1} = \frac{(25\,\text{W})}{(3.35 \times 10^{-19}\,\text{J})} = 7.4 \times 10^{19}$. (1)

4 You will find the argument on page 72. (1)

The longest wavelength corresponds to $hf = \frac{hc}{\lambda} = \phi$. (1)

Substituting gives $\phi = \frac{(6.6 \times 10^{-34}\,\text{J s})(3.0 \times 10^{8}\,\text{m s}^{-1})}{(6.5 \times 10^{-7}\,\text{m})} = 3.0 \times 10^{-19}$ J. (2)

5 The photon energy $E = \frac{hc}{\lambda}$. Substituting gives

$$E = \frac{(6.6 \times 10^{-34}\,\text{J s})(3.0 \times 10^{8}\,\text{m s}^{-1})}{(1.5 \times 10^{-7}\,\text{m})} = 13.2 \times 10^{-18}\,\text{J}. \text{ (1)}$$

The work function of sodium $= 2.5\,\text{eV} = 4.0 \times 10^{-19}\,\text{J}$. (1)

Rearranging Einstein's equation $E = \phi + E_k$ gives $E_k = E - \phi$. (1)

$E_k = 13.2 \times 10^{-19}\,\text{J} - 4.0 \times 10^{-19}\,\text{J} = 9.2 \times 10^{-19}\,\text{J}$. (1)

6 For the LED to emit light the photon energy of the light must be less than the p.d. across the LED multiplied by the electronic charge. (1)

The photon energy $E = \dfrac{hc}{\lambda}$ $E = \dfrac{(6.6 \times 10^{-34}\,\text{J s})(3.0 \times 10^{8}\,\text{m s}^{-1})}{(6.5 \times 10^{-7}\,\text{m})} = 3.0 \times 10^{-19}\,\text{J}$. (1)

Convert E to eV. $E = \dfrac{(3.0 \times 10^{-19}\,\text{J})}{(1.60 \times 10^{-19}\,\text{J eV}^{-1})} = 1.9\,\text{eV}$. (1)

The data is consistent as each electron gains 2.1 eV going through the LED.

7 The energy level diagram shows the arrow as (i), (ii) and (iii). (3)

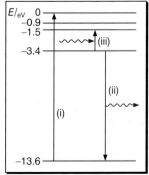

DE BROGLIE, DÖPPLER AND HUBBLE (page 79)

1 For electrons behaving as particles you might include: a beam of electrons in a cathode ray tube, electrons casting a shadow in a Maltese Cross tube, photoelectrons in the photoelectric effect, electrons in the thermionic effect (electrons 'boiled off' a heated filament). (2)

For electrons behaving as waves you might include: the electron diffraction tube, the electron microscope forming an image, electron wave functions in atoms. (2)

2 Rearranging $p = \dfrac{h}{\lambda}$ gives $\lambda = \dfrac{h}{p}$. (1)

Substituting gives $= \dfrac{(6.6 \times 10^{-34}\,\text{J s})}{(4.5 \times 10^{-24}\,\text{N s})} = 1.5 \times 10^{-10}\,\text{m}$. (1)

3 For diffraction effects to be observed, the slit separation must be comparable to the de Broglie wavelength of the electrons. (1) The separation of planes of atoms in a crystal is comparable to the wavelength but 0.5 mm is not. (1)

4 For the bullet $p = mv = (0.015\,\text{kg})(350\,\text{m s}^{-1}) = 5.25\,\text{N s}$. (1)

Check yourself answers

Rearranging $p = \frac{h}{\lambda}$ gives $\lambda = \frac{h}{p}$. (1)

Substituting gives $= \frac{(6.6 \times 10^{-34}\,\text{J s})}{(5.25\,\text{N s})} = 1.3 \times 10^{-34}\,\text{m}$. (1)

This value is 1020 times smaller than the size of a nucleus; far too small to observe diffraction effects. (1)

5 The formula required is $\frac{\Delta f}{f} = \frac{v}{c}$, rearranged to give $v = c\frac{\Delta f}{f}$. (1)

Velocity of recession $v = \frac{(3.0 \times 10^{8}\,\text{m s}^{-1})(3.0 \times 10^{12}\,\text{Hz})}{(5.0 \times 10^{14}\,\text{Hz})}$, which equals $1.8 \times 10^{6}\,\text{m s}^{-1}$. (1)

RADIOACTIVITY (page 86)

1 Phosphorus has an atomic number (Z) of 15, so 15 protons. (1) The mass number is 32, so the number of neutrons is (32 − 15) = 17. (1)

2 To find the number of α-particles. Only α-particles change the mass number. The total change in mass number is (238 − 206) = 32, which corresponds to 8 α-particles each of mass number 4. (1)
8 α-particles, each of charge 2 e, will reduce the total charge by 16 e. (1)
The actual drop in charge is (92 − 82) = 10 e. (1) 6 β-particles will reduce the drop in charge to 10 e. Therefore the total process involves the emission of 8 α-particles and 6 β-particles. (1)

3 β-decay results in the daughter nucleus having an increase in Z of +1 and no change in A. (2) Therefore, $^{60}_{27}\text{Co}$ $^{60}_{28}\text{Ni} + ^{0}_{-1}\beta + 2\gamma$. (2)

4 All three isotopes are of oxygen, which has an atomic number Z of 8. Therefore all three isotopes have 8 electrons and 8 protons. (2) Each isotope has a different number of neutrons = (A − 8). Oxygen-16 has 8, oxygen-17 has 9 and oxygen-18 has 10 neutrons. (2)

5 The half-life of an isotope is the time taken for the activity of the sample to halve in value. (1)
75 h is $(\frac{75}{15}) = 5$ half-lives of the sodium-24. (1) The count rate will have fallen by a factor of 2 five times: a factor of $2^5 = 32$. (1) Therefore count rate after 75 h is $\frac{(4000\,\text{Bq})}{32} = 125$ Bq. (1)

6 Choose any two count rates differing by a factor of 2 and read off the time interval over which the count rate has fallen by a factor of 2. For example, the count rate has fallen from $800\,\text{s}^{-1}$ to $400\,\text{s}^{-1}$ in 150 s. (3)

7 Read the conclusion from Rutherford's experiment on page 82. (3) Large values of Z and A are desirable because:
Large Z means a large positive charge which repels the α-particle strongly and ensures elastic scattering. (1)
Large A means that the target nucleus will have little recoil energy when bombarded by the α-particle. (1)

Check yourself answers

1 For the proton the charge is +1 and the baryon number is +1. (1)
For (u u d) charge $= \frac{2}{3} + \frac{2}{3} - \frac{1}{3} = +1$, which is correct (1) and baryon number $= \frac{1}{3} + \frac{1}{3} + \frac{1}{3} = +1$, also correct. (1)

2 The neutron is (u d d). (1) β-decay may be represented as a neutron turning into a proton $+\beta^- + \bar{\nu}_e$. (1) Comparing the quark compositions of p and n, a d-quark has changed into a u-quark. (1)

3 The rest energy of an electron or a positron is 0.51 MeV. (1) Therefore, to create an electron–positron pair will require at least 1.02 MeV. (1) When a low-energy electron–positron pair annihilates the total γ-ray energy is 1.02 MeV shared between the two γ-rays. (1) To conserve momentum the two γ-rays must have equal energy and go off in opposite directions. (1)

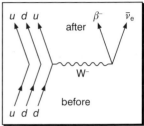

4 The rest mass of a proton is 1 u equivalent to 930 MeV. (1)
To create the extra proton–antiproton pair will require 1860 MeV. (1)
The fraction of energy used to create the pair $= 1.86 \frac{GeV}{10}$ GeV. (1)
Fraction = 0.19. (1)

5 For the π^0 we require charge and baryon number both to be zero. (1) This could be achieved by a π^0 being (u $\bar{u}$). (1)
For the π^+ we require a charge of +1 and a baryon number of zero. (1) This could be achieved by π^+ being (u $\bar{d}$). (1)

6 Apply the conservation laws:

charge	−1	=	−1	+0 +0	✔
lepton number	+1	=	+1	−1 +1	✔
baryon number	0	=	0	+0 +0	✔

All three correct so this decay will happen

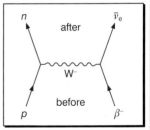

7 The diagram represents a high-energy electron colliding with a proton. (1) The electron converts the proton into a neutron (1) and is concerted into an electron-neutrino. (1) The intermediate exchange particle is a W⁻. (1)

1 The formula required is $E = mc\Delta\theta$, where $\Delta\theta = (100 - 20)\,^\circ C = 80\,^\circ C$. (1)
Substituting gives $E = (0.50\,kg)(4200\,J\,kg^{-1}\,K^{-1})(80\,^\circ C) = 170\,kJ$. (1)

2 The formula power $= \frac{energy}{time}$ is rearranged to give time $= \frac{E}{P}$. (1) Substituting gives time taken $= \frac{(170\,000\,J)}{(2000\,W)} = 85\,s$. (1)

Check yourself answers

3 Energy transferred to water in 1 min = (2000 W)(60 s) = 120 kJ. (1)
The formula $E = mL$ is rearranged to give $m = \frac{E}{L}$. (1)

Substituting gives $m = \dfrac{(1.2 \times 10^5 \text{ J})}{(2.2 \times 10^6 \text{ J kg}^{-1})}$ = $\underline{0.055\text{ kg}}$. (1)

4 To cool 1 kg of water by 8°C requires the removal of (4200 J kg–1)(8 K) or 34 kJ of energy from the water. Remember that a change of 1 K is the same change as a change of 1°C. (1) To freeze 1 kg of water requires the removal of 330 kJ of energy from the water. (1) Freezing requires the removal of ten times as much energy (1) and so will take much longer, assuming the freezer is maintained at a constant temperature. (1)

5 **(a)** (1)   **(b)** (1)   **(c)** (1)

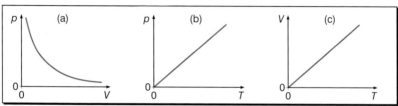

6 For a fixed mass of gas $\frac{pV}{T}$ is constant. (1)

Initially $\frac{pV}{T} = \dfrac{(100\text{ kPa})(5.03\text{ m}^3)}{(290\text{ K})}$ = 1700 J K^{-1}. (1)

Finally, $\frac{pV}{T} = 1700\text{ J K}^{-1} = \dfrac{(30\text{ kPa})V}{(240\text{ K})}$.

Rearranging gives $V = \dfrac{(1700\text{ J K}^{-1})(240\text{ K})}{(30\text{ kPa})}$ = 14 m^3. (1)

7 At constant temperature the speed of the particles does not change. (1) As the volume increases the time between collisions with the container walls increases and so the number of collisions per second with the walls decreases. (1) Fewer collisions with the walls per second per unit area leads to a lower pressure. (1)

8 Mean kinetic energy of helium atom $= \frac{3}{2}kT = \frac{3}{2}(1.4 \times 10^{-23}\text{ J K}^{-1})(300\text{ K})$
$= 6.3 \times 10^{-21}$ J. (2)
Kinetic energy $= \frac{1}{2}m\overline{c^2}$, which rearranges to give $\sqrt{\overline{c^2}} = \sqrt{2\frac{E_k}{m}}$. (1)
Substituting gives $\sqrt{\overline{c^2}} = \sqrt{\dfrac{(12.6 \times 10^{-21}\text{ J})}{(6.7 \times 10^{-27}\text{ kg})}}$ = 1400 m s^{-1}. (1)

INFORMATION AND DATA EXCHANGE (page 101)

1 A p.d. of zero is a '0' and of V is a '1'. This train of pulses reads as 010110001010. (2)

2 Points you might make include that digital information is easier to store, is easier to edit and manipulate (information from a body scan can be

Check yourself answers

manipulated to give different sections through the body), can be transmitted more accurately as it can be cleaned more easily to remove noise and interference, is compatible with modern computer systems. Any four points. (4)

3 The analogue-to-digital converter changes the analogue output of the sensor into a digital format. The processor edits the information into a form suitable for storage or display. The memory stores the information the display outputs the information via a printer, VDU, etc. (4) Diagram; see page 97. (1)

4 Number of bits of information in one picture is 2.1×10^6 pixel 16 bit pixel^{-1} $= 3.36 \times 10^7$ bit. (1)

1 byte = 8 bit so 3.36×10^7 bit $= 4.2 \times 10^6$ byte = 4.2 Mbyte. (2)

5 Each sample has 16 bits and there are 4×10^4 samples taken per s. (1)
Therefore number of bit s^{-1} = $14 \times (4 \times 10^4)$ = 640 kbit. (1)

6 The maximum bit rate = 5×10^9 bit s^{-1} and one TV channel requires 25 Mbit s^{-1}. (1) Therefore number of channels $= \dfrac{(5 \times 10^9)}{(2.5 \times 10^7)} = 200$. (1) In practice, fewer channels could be transmitted because of the need to provide spaces and bits to identify the channel. (1)

7 Advantages you could include are: greater bandwidth and therefore more information per second, freedom from electrical interference and therefore fewer faults in transmission, smaller attenuation and therefore greater distance between repeaters, smaller physical size, etc. (1 + 1) twice.

Quantity	Unit	Abbreviation	Other units for same quantity
mass	kilogram	kg	[SI base unit]
length	metre	m	[SI base unit]
density	kilogram per metre^3	$kg\,m^{-3}$	
time	second	s	[SI base unit]
speed, velocity	metre per second	$m\,s^{-1}$	
acceleration	metre per second^2	$m\,s^{-2}$	
force	newton	N	$kg\,m\,s^{-2}$
weight	newton	N	
pressure	pascal	Pa	$N\,m^{-2}$
stress	pascal	Pa	$N\,m^{-2}$
strain	no unit		(a ratio of two lengths)
moment of a force	newton metre	$N\,m$	
energy	joule	J	$1\ eV = 1.6 \times 10^{-19}\,J$
power	watt	W	$J\,s^{-1}$
momentum	newton second	$N\,s$	$kg\,m\,s^{-1}$
electric current	ampere	A	[SI base unit]
electric charge	coulomb	C	$A\,s$
electric p.d.	volt	V	$J\,C^{-1}$
electrical resistance	ohm	Ω	$V\,A^{-1}$
resistivity	ohm metre	$\Omega\,m$	
electrical conductance	siemens	S	Ω^{-1}
conductivity	siemens per metre	$S\,m^{-1}$	
magnetic flux density	tesla	T	
frequency	hertz	Hz	s^{-1}
wavelength	metre	m	
refractive index	no unit		(a ratio of two speeds)
temperature	degree celsius	°C	
absolute temperature	kelvin	K	[SI base unit]
amount of substance	mole	mol	[SI base unit]
radioactivity	becquerel	Bq	s^{-1}

Here is a table of electrical circuit symbols. You need to know these to both draw and interpret circuit diagrams in questions.

Name of component	Symbol
junction of two conductors	
conductors crossing but not connected	
switch (normally open)	
terminals	
cell	
battery or group of cells	
voltmeter	
ammeter	
fixed resistor	
indicator or lamp	
light-dependent resistor (LDR)	
thermistor	
potentiometer or voltage divider	
semiconductor diode	
light-emitting diode	

You need to know the formulae on this list as they will not be given on question papers or formulae lists in examinations.

density = $\dfrac{\text{mass}}{\text{volume}}$	$\rho = \dfrac{m}{v}$
speed = $\dfrac{\text{distance}}{\text{time taken}}$	$v = \dfrac{s}{t}$
acceleration = $\dfrac{\text{change in velocity}}{\text{time taken}}$	$a = \dfrac{\Delta v}{\Delta t} = \dfrac{(v - u)}{t}$
force = mass × acceleration	$F = ma$
momentum = mass × velocity	$p = mv$
work done = force × distance moved	$E = Fx$
power = work done/time taken	$P = \dfrac{E}{t}$
weight = mass × gravitational field strength	$W = mg$
kinetic energy = $\frac{1}{2}$ × mass × speed^2	$E_k = \frac{1}{2} mv^2$
change in gravitational potential energy = mass × gravitational field strength × change in height	$\Delta E_P = mg\Delta h$
pressure = $\dfrac{\text{force}}{\text{area}}$	$p = \dfrac{F}{A}$
pressure × volume = amount (in mol) × molar gas constant × absolute temperature	$pV = nRT$
charge = current × time	$\Delta q = I\Delta t$
potential difference = current × resistance	$V = IR$
electrical power = potential difference × current	$P = VI$
potential difference = $\dfrac{\text{energy}}{\text{charge}}$	$V = \dfrac{E}{q}$
energy = potential difference × current × time	$E = VIt$
resistance = $\dfrac{\text{resistivity} \times \text{length}}{\text{area of cross-section}}$	$R = \dfrac{\rho l}{A}$
wave speed = frequency × wavelength	$v = f\lambda$